Open Science: the Very Idea

Frank Miedema

Open Science: the Very Idea

 Springer

Frank Miedema
UMC Utrecht
Utrecht University
Utrecht, The Netherlands

ISBN 978-94-024-2117-0   ISBN 978-94-024-2115-6 (eBook)
https://doi.org/10.1007/978-94-024-2115-6

This Springer imprint is published by the registered company Springer Nature B.V.
The registered company address is: Van Godewijckstraat 30, 3311 GX Dordrecht, The Netherlands

For Yuna and Mare,
It's all about your future

Preface

First, I want to thank the readers for starting to read my book. Somehow it got your attention despite the daily tsunami of information, infotainment and entertainment that is thrown at you. Most of us are pressed for time, even in months of lockdown because of COVID-19, so I will keep this introduction short. I will do without the usual clever well-worded academic introduction, referring to Aristotle, Popper and Foucault. I will go straight to the relevant question any reader from academia, public or policy could ask with no need to be ashamed of:

> Why a book about Open Science? It is a hype isn't it? So, is there not a lot already written about it? Ok, it is a big thing in Europe, Australia, and even China it seems, but what about the USA and Canada, and are the Germans really in? I do recall, in the USA during the Obama administrations there was some action of Open Access and Open Data, and academics who started DORA in 2012 in San Francisco.

or

> A book about Open Science? Open Access you mean? Does that start again? Wasn't that the movement that, since the year 2000 had a few starts already, sympathetic but that just didn't take of? Does the EU still believe in it? Really? Why?

In this book I address these questions, without any beating around the bush. It is truly amazing, that a way to do science and research, that for the majority of its practitioners and the public and policy makers makes a lot of sense, and which has been around for quite some time, has not been embraced to become common practice. To answer this question, we have to delve deep into the science of science and research. We have to understand 'the idea of science' that does exist in the plural. We have to analyse why in particular one of these concepts and its corresponding public image has been dominant practically since 1945 and what that has done to science and scientists. That philosophical/sociological idea has been the basis for the ideologic narrative with which science has been internally organized and is being used to claim a unique position, authority and funding for science. With this narrative, the scientific community promised that science would be there to the benefit of society, at least when her autonomy and neutrality are respected. How come that although this legendary image and its narrative by the philosophers, historians and sociologists has no philosophical and timeless foundation, scientists apparently without knowing this demise of their Legend keep using that narrative? It may well be the fear, the insecurity that comes with the awareness that knowledge production in science is based not on a given metaphysical foundation, but rests on a firm social process of a community of inquirers that relentlessly criticize, question, debate what the best knowledge claims are. Knowing very well that the consensus reached may work well but is never absolute and may be replaced by better ones by this same process of inquiry called science. Having said this, we realize that, despite the commonly held views, the 'method' of the 'hard' sciences and that of the 'soft' social science and humanities may not be all that different after all!

In our present-day world of hyper-modernity, where knowledge is everywhere to be found and always contested by some, the process of the production of knowledge cannot be insulated from potential users and interested critical other parties. Clinging to the idea of a unique method for absolute truth and a foundation for science is understandable but a wrong reflex in debates with the public about its problems. Explaining how science really works and produces knowledge would be the best response.

In this book, I and you readers need to be totally frank about science, and we need to be 'biting the bullet' and bringing up several difficult issues. We need to discuss 'therapeutic' interventions required to opening up research and academia for transition to a more open science that works better for the world. You may wonder, 'is the relationship with the public then the problem?' It sure is, since I believe, with many colleagues, that scientists have a moral obligation to engage with the major societal problems and challenges of their time. I may seem very optimistic, but I am not naïve, to think that the practice of Open Science will be a major improvement regarding the relationship between science and society in several critical ways.

In contrast to the critical questions, it appears that since 2016 the idea of Open Science has been adopted by many institutions and governments around the world and it is well possible that we have passed the tipping point of its global breakthrough. Life is never perfect, and as we know Open Science to function properly needs an Open Society, but this requirement is not universally fulfilled and needs the attention of academia. I clearly see an opportunity for a leading role of the Open Science approach of the EU, now the USA has lost a lot of its position as a world leader in science. I have touched upon the problem of geopolitics, for instance in relation to China, the new science superpower, as this is not specific for science but a general problem of democracy it is outside the scope of this book.

The good news is that in the heat of the catastrophic COVID-19 pandemic, with its unprecedented global threat to public health and our socio-economic life, we see the opening up of the different practices of science, publishing, data and biomaterial sharing, and doing research and in real time opening up to public, at national and international levels. It is argued here that science should always be done like that, as Open Science.

My Journey

This book results from my journey in science since the early 1970s. I had the opportunity of being at the bench of biochemistry laboratories of different knowledge institutions: in the Netherlands at The National Institute for Public Health and the Environment, the non-profit Blood Supply Foundation and two University Medical Centres, with a six months 'sabbatical' in 1994 at DNAX Research Institute, Palo Alto, CA. From 1975 on, I was intrigued by the social aspects of science. While in my academic development I discovered the practice of research, in parallel I discovered the science of science and developed an interest as a science observer. Ever since, while being totally engaged in research, at the same time I was a science observer and obsessively studied the many aspects of that social game called science. It is this dual experience and the broad view of science that has led to my writings and actions to improve science of which this book gives witness reporting from both tracks of my journey in science.

Finally, I have provided a broad and, in some cases, an in-depth background analysis of different aspects of science. Importantly, regarding the images of science that still distort our image and practice of science, most writers about the practice of science mention the problem but almost all refrain from such in-depth analysis. Almost all stay even further away from the directly connected discussion of the incentive and rewards systems because that unavoidably opens the black box, or rather the can of worms of academic politics, the game of reputation, power and money. Since, as I argue in Chap. 2, I believe that the obsolete philosophy and public image of science are the major cause of many problems in the practice of science, I went all out to present the major different arguments for the lack of foundation for the methods of empirical positivism, and its analytical foundational philosophy. I also dwell on the analyses of the problems of the practice of science in academia, since I do regard this in many ways an important and necessary step for the transition to Open Science. Finally, there is a lengthy description of the development, from early initiatives to finally the institutional start of Open Science in the research agenda of the European Union. I realize that most of you shall want to read selectively according to your main immediate interests. Therefore, I provide four distinct reading tracks below.

Four Recommended Reading Tracks

0. **For the general concise view of science and society since 1945**
   Chap. 1.
1. **Philosophy and sociology**
   If you are more interested and have read in the philosophical and sociological origins of our current ideas about science, Chaps. 2 and 4 will be serious reading but are highly recommended.
2. **A critique of science**
   If you want to grasp the more recent critical thinking about science, with analyses and arguments in the pre-Open Science time, then go for Chaps. 1, 3 and 6.
3. **New avenues**
   If you don't want the diagnosis, but rather read about attempts and ideas how, by engaging with society, to improve the relationship of science with society, that is to be found in Chaps. 5, 6 and 7.
4. **Transition to Open Science**
   If you want an impression about early actions in the past 20 years and the more recent actions taken to promote Open Science, go to Chaps. 5 and 7, and for some of the local initiatives in Utrecht, I refer to Chap. 6.

Utrecht, The Netherlands   Frank Miedema

Synopsis

Chapter 1 : Science and Society An Overview of the Problem

Science promised to society to contribute to the grand challenges of the United Nations, WHO, the EU agenda and national agendas for change and improvement of our life. It will be discussed how this social contract between science and society has developed since 1945. The first phase from 1945 till 1960 was characterized by autonomy, building on the successes of the natural sciences and engineering in World War II. In the second phase, the late 1960s till approximately 1980, government and the public lost trust and saw the downside of science and technology. The response from politics and the public was a call for societal and political responsible research inspired by broader socio-political developments in society. The third phase from 1980 till 2010 was built on the idea that science and technology would bring economic growth, which should make nations internationally competitive. There was also increasingly room for societal problems related to environment and sustainability, health and well-being. In this approach of the so-called knowledge economy, strong relations with government and the private sector were established characterized by short-term accountability, control from government and funders at the level of project output, using accordingly defined metrics and indicators. This model became firmly and globally institutionalized.

Within science, since 2010 among scientists there is growing frustration, mostly implicit but increasingly explicit disillusion of scientists, regarding governance, agenda setting in relation to the outside world and significant impact of the research. Science fails, it is felt, its promise to society to contribute to the quality of life as the system has adapted to the culture of new public management. Production of robust and significant results is mainly secondary to output relevant for an internal credit system for academic career advancement at the individual level. At the higher organizational level output and impact are focused on positions on international ranking lists which drive highly competitive social systems which results in a widely felt lack of alignment and shared value in the academic community.

Chapter 2: Images of Science: A Reality Check

It will be argued that the dominant form of current academic science is based on ideas and concepts about science and research that date back to philosophy and sociology that was developed since the 1930s. It will be discussed how this philosophy and sociology of science has informed the ideas, myths and ideology about science held by the scientific community and still determines the popular view of science. It is even more amazing when we realize that these ideas are philosophically and sociologically untenable and since the 1970s were declared obsolete by major scholars in these same disciplines. To demonstrate this, I delve deep to discuss the distinct stages that scholars in philosophy, sociology and history of science since 1945 to 2000 have gone through to leave the analytical-positivistic philosophy behind. I will be focusing on developments of their thinking about major topics such as how scientific knowledge is produced, the scientific method, the status of scientific knowledge and the development of our ideas about 'truth' and the relation of our claims to reality. It will appear that the positivistic ideas about science producing absolute truth, about 'the unique scientific method', its formal logical approach and its timeless foundation as a guarantee for our value-free, objective knowledge were not tenable. I took the trouble to go into this deep to show how thoroughly the myth has been demystified in philosophy and sociology of science. You think after these 50 pages I am kicking a dead horse? Not at all! This scientific demystification has unfortunately not reached active scientists. The popular image of science and research is still largely based on that Legend. This is not without consequence as will be shown in Chap. 3. These images of science have shaped and in fact distorted the organizational structures and the interaction between its institutes and disciplines. It also affects the relationship of science with its stakeholders in society, its funders, the many publics private and public and policy makers in government. In short, it is about the growth of knowledge.

It will throughout the book be accompanied by a narrative in which I will take my own intellectual and scientific journey from 1971 as a chemistry student who did a minor in the philosophy of science in academic year 1975–1976. Since then I followed the classical career path of a professional biochemist/ immunologist, as PhD student, post-doc, group leader, department head, director of a small research institute, to finally become dean and board member of a large university medical centre. Going through this professional sequence, I kept a persistent and ever stronger interest in the science of science. It is from the perspective of a true understanding of the practice of science in its various aspects that I will use specific authors a lot, but others much less or even neglect work of many scholars that to specialists in the different fields are considered important but are of little or no relevance for the daily practice of active researchers and most other actors in the field.

Chapter 3: Science in Transition How Science Goes Wrong and What to Do About It

Science in Transition which started in 2013, is a small-scale Dutch initiative. It presented a systems approach, comprised of analyses and suggested actions, based on experience in academia. It was built on writings by early science watchers and most recent theoretical developments in philosophy, history and sociology of science and STS on the practice and politics of science. This chapter will include my personal experiences as one of the four Dutch founders of Science in Transition. I will discuss the message and the various forms of reception over the past 6 years by the different actors in the field, including administrators in university, academic societies and Ministries of Higher Education, Economic Affairs and Public Health but also from leadership in the private sector. I will report on my personal experience of how these myths and ideologies play out in the daily practice of 40 years of biomedical research in policy and decision making in lab meetings, at departments, at grant review committees of funders and in the Boardrooms and the offices of Deans, Vice Chancellors and Rectors.

It has in the previous chapters become clear that the ideology and ideals that we are brought up with are not valid and are not practiced; despite that even in 2020 they are still somehow 'believed' by most scientists and even by many science watchers and journalists and used in political correct rhetoric and policy making by science's leadership. In that way these ideologies and beliefs mostly implicitly but sometimes even explicitly determine debates regarding the internal policy of science and science policy in the public arena. These include all-time classic themes like the uniqueness of science compared to any other societal activity; ethical superiority of science and scientists based on Mertonian norms; the vocational disinterested search for truth, autonomy; values and moral (political) neutrality; dominance of internal epistemic values; and unpredictability regards impact. These ideas have influenced debates about the ideal and hegemony of natural science; the hierarchy of basic over applied science; theoretical over technological research; and at a higher level in academic institutions and at the funders the widely held supremacy of STEM over SSH. This has directly determined the attitudes of scientists in the interaction with peers within the field, but also shaped the politics of science within science but also with policy makers and stakeholders from the public and private sector and interactions with popular media.

Science it was concluded was suboptimal because of growing problems with the quality and reproducibility of its published products due to failing quality control at several levels. Because of too little interactions with society during the phases of agenda setting and the actual process of knowledge production, its societal impact was limited which also relates to the lack of inclusiveness, multidisciplinarity and diversity in academia. Production of robust and significant results aiming at real-world problems is mainly secondary to academic output relevant for an internally driven incentive and reward system steering for academic career advancement at the individual level. Similarly, at the higher organizational and national level this reward system is skewed to types of output and impact focused on positions on international ranking lists. This incentive and reward system, with flawed use of metrics, drives a hyper-competitive social game in academia which results in a widely felt lack of alignment and little shared value in the academic community. Empirical data, most of it from within science and academia, showing these problems in different academic disciplines, countries and continents are published on a weekly basis since 2014. These critiques focus on the practices of scholarly publishing including Open Access and open data and, the adverse effects of the incentive and reward system. Images, ideologies and politics of science were exposed that insulate academia and science from society and its stakeholders, which distort the research agenda and subsequentially its societal and economic impact.

Chapter 4: Science and Society: Pragmatism by Default

To rethink the relation between science and society and its current problems authorative scholars in the USA and Europe, but also around the globe, have since 1980 implicitly and increasingly explicitly gone back to the ideas of American pragmatism. Pragmatism as conceived by its founders Peirce, James and Dewey is known for its distinct philosophy/sociology of science and political theory. They argued that philosophy should not focus on theoretical esoteric problems with hair-splitting abstract debates of no interest to scientists because unrelated to their practice and problems in the real world. In a realistic philosophy of science, they did not accept foundationalism, dismissed the myth of given eternal principles, the unique 'scientific method', absolute truths or let alone a unifying theory. They saw science as a plural, thoroughly social activity that has to be directed to real-world problems and subsequent interventions and action. 'Truth' in their sense was related to the potential and possible impact of the proposition when turned in to action. Knowledge claims were regarded per definition a product of the community of inquirers, fallible and through continuous testing in action were to be improved. Until 1950, this was the most influential intellectual movement in the USA, but with very little impact in Europe. Because of the dominance of the analytic positivistic approach to the philosophy of science, after 1950 it lost its standing. After the demise of analytical philosophy, in the 1980s, there was a resurgence of pragmatism led by several so-called new or neo-pragmatists. Influential philosophers like Stephen Toulmin, Hillary Putnam and Philip Kitcher coming from the tradition of analytic philosophy have written about their gradual conversion to pragmatism, for which in the early days they were frowned upon by esteemed colleagues. This movement gained traction first in the USA, in particular through works of Bernstein, Toulmin, Rorty, Putnam and Hacking, but also gained influence in Europe, early on through the works of Apel, Habermas and later Latour.

Chapter 5: Science in Social Contexts

Gradually since 1990 a growing number of critical analyses from within science have published on how science was organized as a system and discussing its problems, despite, or paradoxically because the growing size of its endeavour and its growing yearly output. Because of lack of openness with regard to sharing results of research, such as publications and data but in fact of all sorts of other products, science is felt by many to disappoint with respect to its societal impact, its contribution to the major problems humanity is facing in the current times. With the financial crisis, in analogy, also the crisis of the academic system as described in Chap. 3 was exposed and it seemed that similar systemic neoliberal economic mechanisms operated in these at first sight seemingly different industries. Most of these critiques appeared with increasing frequency since 2014 in formal scientific magazines and social media and reached the leadership of universities, government and funders. This awareness and support for the development of new ways of doing science, mostly intuitively and implicitly, but sometimes explicitly, is motivated by pragmatism aiming for societal progress and contribution to the good life.

To get to this next level we need the critical reflection on the practice of science as done in previous chapters in order to make systemic changes to several critical parts of the knowledge production chain. I will discuss the different analyses of interactions between science and society, in the social and political contexts with publics and politics that show where and how we could improve. The opening up of science and academia in matters of problem choice, data sharing and evaluation of research together with stakeholders from outside academia will help to increase the impact of science on society. It ideally should promote equality, inclusion and diversity of the research agendas. This, I will argue, requires an Open Society with Deweyan democracy and safe spaces for deliberations where a diversity of publics and their problems can be heard. In this transition we have to pay close and continuous attention to the many effects of power executed by agents in society and science that we know can distort these 'ideal deliberations' and undermine the ethics of these communications and possibly threaten the autonomy and freedom of research.

Chapter 6: Science in Transition Reduced to Practice

In the true spirit of Dewey and pragmatism, knowledge, insights and experience have to be translated into interventions and actions. Only when knowledge is 'reduced to practice', its social robustness and value will be determined. In light of the conclusions of the previous chapter, to be able to have more impact and to hold up our promise to society, we have to reflect who our science is organized and how it could be improved. From these reflections, several interventions in the practice of research have been proposed. When we, the Science in Transition Team, started to make public our critical accounts of the practice of science, I was 'friendly advised' by influential older scientists to first clean up the mess in my own institution, instead of pointing to others and to the system. As a matter of fact, that is what we have been doing at University Medical Centre Utrecht (UMC Utrecht) since 2009. In this chapter I present a brief outline of our actions 'on the ground' in UMC Utrecht, their receptions and some early actions to promote these activities abroad.

Chapter 7: Transition to Open Science

Many initiatives addressing different types of problems of the practice of science and research have been described or cited. Some were one-issue local actions, some took a broader approach at the national and some at EU level. Some stayed on, others faded after a few years. Many of the issues addressed by these movements and initiatives were part of the system of science and appeared to be systemically interdependent. This is how they converged and precipitated in the movement of Open Science, somewhere at the beginning of the second decade of this century. I discuss the major move that was made since 2015 in the EU to embrace the Open Science practice as the way science and research are being done in Europe. This elicited tensions at first foremost relate to uncertainty regarding scholarly publishing, of how and where we publish open access. But also, with respect to what immediate sharing of data and results in daily practice of researchers means, how we value and give credit for papers and published data sets. It thus poses the question of recognition and rewards, how, if at all, we must compare incomparable academic work, how we get credit and build reputations in this new open practice of science.

It is indeed believed that Open Science with its practice of responsible science will be a major contribution to address the dominant problems in science that we have analysed thus far, or at least will help to mitigate them. Open Science holds a promise to take science to the next phase as outlined in the previous chapters. That is not a romantic naive longing for the science that once was. It will be a truly novel, but realistic way of doing scientific inquiry according to the pragmatic narrative pointed out.

I will conclude this chapter reporting some of my firsthand experiences, in Brussels and during visits to several EU member states in the course of a Mutual Learning Exercise, but also encounters in North America, South-East Asia and South Africa where we in the past years have discussed Open Science. We know science and scholarship have many forms and flavours and that wherever you go around the globe, there is not one scientific community. For me discussing the Transition to Open Science in the past four years was really a Learning Exercise, an amazing, mostly encouraging, but many times quite shocking, humbling and even sometimes saddening experience.

The Transition to Open Science, as can be anticipated from the analyses above, will not be trivial. The recent discussions have already shown that the transition to Open Science, even in different EU Member States, is a very different thing because of specific national, cultural and academic contexts.

Acknowledgements

The plan for writing this book originates from 2013, the time when we started Science in Transition. I had been writing essays and book reviews about science and had published some of them in English with Amsterdam University Press in 2012. With the start of the Science in Transition movement, we wrote and talked continuously about science in the broadest possible sense. During my seminars since then philosophy, sociology and politics of science were admixed with personal experiences that illuminated the analyses of how science works, has gone wrong and what to do about it. Rinze Benedictus, MSc in medical biology and long-term science writer, was from the start involved in Science in Transition and before that as staff member had been working with us on the *UMCUtrecht 3.0 Strategy 2010–2015*. From his dual experience he recognized the journalistic power of mixing theoretical analysis with personal experiences and urged me to write it up or, given my lack of free time, at least contract a professional science writer who could do that with/for me. An important 'boost' to Rinze's 'priming' came unexpectedly from Daniel Sarewitz. As I describe in Chap. 6, during a breakfast meeting with Dan Sarewitz and Paul Wouters, in January 2017 at Washington DC, I told Dan the story of the organizational intervention at UMC Utrecht which prompted the change in our way of research evaluation later on. Sarewitz, being a successful writer himself, immediately said that I should write it up since it demonstrates how evaluations ('metrics' if you like) follow strategy. Against Rinze's advice, I waited until March 2019 when I stepped down from the Executive Board of UMC Utrecht, trusting that the story I had to tell would still be relevant.

I am most grateful to the Board of UMC Utrecht and the Board of Utrecht University to jointly appoint me, per March 1, 2019, as chairperson of the UU Open Science Program which was an additional inspiration for me to write this book.

During the process of writing, but in fact since 1971, I was encouraged by frequent discussions with Siebren Miedema, my brother who is by four years my senior and emeritus professor in Educational Foundations and Religious Education.

The manuscript or parts of it have been read by Sarah de Rijcke, René von Schomberg, Anja Smit, Huub Dijstelbloem, Jerome Ravetz, Frank Huisman, Wijnand Mijnhardt, Gerard de Vries, Siebren Miedema, Rinze Benedictus and

Susanne van Weelden. The members of the Utrecht University Open Science Team, Judith de Haan, Tom Peijster, Suzanna Bloem and Sicco de Knecht I thank for daily discussions and support while I worked on the book.

I am indebted to Paul Wouters, Melanie Peters, Albert Meijer, Kees Schuyt, Floris Cohen, Stuart Blume, Barend van de Meulen, Paul Wouters, Hans Chang, Bas van Bavel, Ulli Dirnagl, Daniel Sarewitz, John Ioannidis, Steven Goodman, David Moher, members of the UU Ethics Institute (especially Marcus Duwell and Joel Anderson), Patrick Bossuyt, Lex Bouter, Joeri Tijdink, staff members of the Athena Institute VU (especially Jacqueline Broerse), the WTMC Graduate School (especially Anne Beaulieu), GEWINA (especially Martijn van de Meer), Margriet Schneider, Wim Kremer, Berent Prakken, Wilfrid Opheij, Anna Ridderinkhof and Arjan Miedema for support and critical constructive discussions and advice.

Contents

Chapter 1
Science and Society an Overview of the Problem

Abstract Science in the recent past promised to society to contribute to the grand challenges of the United Nations, UNESCO, WHO, the EU agenda and national agendas for change and improvement of our life, the human condition. In this chapter it will be briefly discussed how this social contract between science and society has developed since 1945. In the context of this book I distinguish three time periods, but I do realize slightly different time periods may be preferred, based on the perspective taken. The first phase from 1945 till 1960 is characterized by autonomy, building on the successes of the natural sciences and engineering in World War II. In the second phase, the late sixties till approximately 1980, government and the public lost trust and saw the downside of science and technology. The response from politics and the public was to call for societal and political responsible research inspired by broader socio-political developments in society. The third phase from 1990 till 2010 was one of renewed enthusiasm and hope that science and technology would bring economic growth, which should make nations internationally competitive. There increasingly was also room for societal problems related to environment and sustainability, health and well-being. In this approach of the so-called knowledge economy, with the world-wide embracing of neoliberal politics, strong relations with government and the private sector were established. This was accompanied by short-term accountability, control from government and funders at the level of project output, using accordingly defined metrics and indicators. Because of this, this model became firmly and globally institutionalized.

It is beyond any doubt that knowledge and innovation are more than ever critically needed to address the current global problems of society that affect our lives, that scientific investigation and the huge public investments involved, could have and must have significantly more impact. This is echoed and pursued by governments, NGO's and others in recent reports and strategic plans (UN, EU, UNESCO, IS7). Do we need such frequent calls upon the responsibility of the science community? Aren't they engaged? Do I need to write this book? Indeed, the calls upon science are timely as ever before, one must say. It is timely and rather urgent for quite different reasons. The factors I will discuss here relate to how science and academia have become and, in many ways still are organized and how this affects and distorts

F. Miedema, *Open Science: the Very Idea*,
https://doi.org/10.1007/978-94-024-2115-6_1

productive interactions between science and society. From the perspective of society this results in suboptimal agenda setting and thus suboptimal return on investment in science's contribution to the major societal issues of our time and age. Before doing so, I will, like virtually every writer about science, scientific inquiry and academia, make it very clear, that science[1] has produced and is producing many important results. The 'hard' natural sciences in particular since the industrial revolution have had enormous impact on the human condition, on the quality of our lives. Scientific investigation and the community of its practitioners had in the seventeenth century gone through a critical change which enabled production of solid and practical knowledge that could be tested and certified (Cohen, 2010; Shapin, 1996). This relates in particular to the various fields of natural sciences like physics, chemistry and engineering. In the twentieth century it has been followed with major progress in especially biomedical research and the geosciences, but also research in psychology, sociology, economy, history and ethics and philosophy have irreversibly changed our lives. New sources of energy, transport and communication, availability of clean water, improvement of general public health mainly through novel measures of hygiene, vaccination and antibiotics, improvement and efficiency of industrial production have impacted our material quality of life. Despite the criticism and the mistrust in science, especially when it brings insights with potentially socially or economically unpleasant consequences, the way scientific knowledge is produced makes it the best institution we have to increase our understanding of the world we live in and understanding ourselves and our life.

Science is a community of peers that puts new findings and claims to the test, that purges and filters it to become robust, reliable, objective and trustworthy knowledge that can guide our actions. At the same time, it keeps producing new knowledge that may, if it survives the reliability tests, replace older beliefs by which mechanism our knowledge growths. This is most accessibly explained by John Ziman (1978, 2000).

Having said this, it does not mean that the organisation of science as it has developed in the past 70 years into an international institution cannot be improved to better serve the needs of the various societies and publics around the globe. It is exactly this notion that I am concerned with. It is about the question how knowledge grows. I will discuss which and whose values and ideas about science and society are involved when we determine its excellence and potential impact and how this determines the research agenda through funding and investments decisions. As the growth of knowledge is not autonomous, not random nor guided by the legendary 'invisible hand', it must be possible to improve the impact of science by cognizant governance to aim at better alignment of the research enterprise with our major national and global societal needs.

---

[1] I will use 'science' or 'the sciences" when I mean to say something about research in all academic disciplines which comprise natural, life science, engineering, the social sciences and scholarship in the humanities. I will use 'science' also when I talk about the total of the academic institutional system of knowledge production.

John Dewey in 1948 concludes in his Introduction to the reprint of his *Reconstruction in Philosophy* (Dewey, 1948), that science had freed us up from religion and 'was *regarded a deliberate assault on the morals that were tied up with the religion of Western Europe(p xii),… but the world and rationality of the natural and technical sciences had deeply entered human daily life (p xiii)'*. The natural sciences have thus entered the domain which was initially '*not only the domain of religious belief and practices, but of virtually all institutions that have been established before the rise of modern science (p xiii)'*. By doing so, he concluded that the original compromise to keep these fields separated, to keep science insulated, had failed. He called for systematic deep philosophical analysis of how that has come about, how it had distorted the old institutions and how we should deal with the significant novel moral issues that come with it. Despite all the technological benefits '*this had not resulted in a world with more security, peace, better governance and higher moral standards*'. This clear critique of the adverse effects of modernity was a broad theme in the thinking of American pragmatism. The use of nuclear weapons, which had happened just 3 years before Dewey wrote these lines, was for many, including leading physicists as Einstein the reason to critically reflect on the societal impact and responsibility of science. Dewey: 'The *development of scientific inquiry is immature; it has not as yet got beyond the physical and physiological aspects of human concerns, interests and subject matters. The institutional conditions into which it enters, and which determine its human consequences have not yet been subjected to any serious, systematic inquiry worthy of being designated scientific'* (p xv) (Dewey, 1948).

I leave it to the reader to reflect on the relevance of these observations for our times. Obviously, science which formally in the US National Science Foundation (NSF) only since 1958 next to the natural sciences includes the social sciences, humanities and engineering, has grown and matured and has become a major global factor in virtually all aspects and domains of public, private and social human life. The relationship between science and society has therefore even become more intricate, more complicated and at the same time more critical regarding the major societal challenges we are facing in the third decade of the twenty-first century.

1.1 A Call for Action

The very week in October 2019, that I started the actual writing of this chapter, *Nature* featured an Editorial on the massive and sad waste in the global food supply system evidenced by research. In the same issue a major big data research paper on mortality in very young children showing global improvement, but still great inequalities in preventable mortality between different geographical regions. In the accompanying comment Michelle Bachelet, a former president of Chile and paediatrician with hands on experience of this problem in that country, argues for an integrated research approach to understand the causes in terms not simply of access to health care, but of 'broader ills: poverty, disempowerment, discrimination and injustice'. 'Hard data', she writes, 'must be followed up by action across a whole spectrum of government and society' (Bachelet, 2019). In that same issue of *Nature,* Diane Coyle, well known from *'The Economics of Enough'*, published a review of three recent books on economics by prominent authors and closes of with a couple of very gloomy lines: *'as Soros asserts that intellectual framework of economics must adapt to a world ever more removed from a focus on individual choices. This trend is under way in economics, but a radical rethink is unlikely there: the incentives of academia encourage conservatism and incremental progress".* She continues: *'Better metrics and theories will not be enough to create a sustainable economic and social model. Or they could- but only if they convince policymakers and the public to act differently'.* As the final blow she adds: *'The future of capitalism is out of the hands of those who spend their time thinking about it'* (Coyle, 2019). That same week in an editorial in *Science,* Ian Boyd chairman of the UK Research Integrity Office, reflects with a strong sense of urgency on the failing interaction between scientists and government and politics (Boyd, 2019). Science should engage more with government and the public debates and let itself not be put in a box with a tight lid on it and being manipulated to become yet another 'money grabbing vested interest'. *'Advocacy is the surest and most rapid way to achieve such an effect'*, he argues *'although science should not be captured by normal politics'.*

These quotes all argue, although in slightly different ways, that science, academic research should aim to have an effect in the real world. As I have argued: *'A paper in Nature does not cure patients'.* It does not change the life expectancy of young children, global socioeconomic systems and policies, the politics, logistics and trade of food, unless it is put into practice, translated to actions to change the condition of those whose quality of life is affected by the problem under study. Science, these writers in October 2019 say, in order to contribute and impact on society, has to connect with the publics, the stakeholders out there who have an interest, an expectation getting a problem solved, having their lives improved. It must result in actions, that then will be put to the test in practice. In this pick of weekend reading of October 19 and 20, 2019, the issues of socioeconomic inequality, public health, societal injustice and food waste are featured, but we also know of grand challenges like climate change; the transition to fossil-free energy; the threats to democracy and its institutions, which includes threats to science by populism and nationalism. In the twenty-first century, challenges are complex, mostly non-linear, which needs a different approach compared to most of the science done before (Beck, 1992; Nowotny, 2016). Science, to be effective must be much more mission-oriented, inclusive, truly multidisciplinary. It should drive not mainly for economic and technological impact, but should also target public and social needs and keep in mind that technology for many problems is not the only solution as it reception by the public will often be poorly understood. Social sciences and humanities (SSH) have to engage since major issues in modern life are in the social and political domain where SSH including economy and political sciences have a lot to offer. The UN has defined seventeen Sustainable Development Goals (SDGs) that must be taken on through science and innovation.[2] To address these goals, breakthrough

---

[2] https://www.un.org/sustainabledevelopment/sustainable-development-goals/

knowledge and solutions will not only come from the natural, biomedical sciences and engineering (STEM). Major issues in these domains relate to problems that need to be investigated in a truly integrated way by researchers from STEM and SSH.

1.2   The Social Contract for Science, What's the Problem?

Why do these authors feel that they have to make these strong pledges to science and the academic community to not only investigate and publish, but to take their academic results one stage further and engage throughout with the relevant agents, policy makers and publics and citizens to which the societal issues matter? What's wrong here? Isn't it, that almost every website of universities and academic institutions around the globe says that its mission, ambitions and strategy are to contribute to the quality of life by excellent research and teaching? That even in most cases explicit societal themes and targets have been chosen in agreement with the UN SDGs?

From the perspective of the public, policymakers, charities and public and private funders one would indeed trust and expect that academic science would be fully geared towards maximal and optimal impact to address and alleviate conditions that interfere with the good life and to address human needs at whatever level in society: personal, structural or political. That is why we think large amounts of public tax money are invested, or to state it more realistic, are being spent.

The relationship, however, between governmental and charity funders and the researchers which have in the previous century become organised in academic institutions and governmental agencies is not that simple. The study of the history of science and its institutions in particular from World War 2 (WW2) on, shows that he connection and interaction between science and society is quite complex and that the aims of the scientific community on the one hand and government and the public on the other, are not always well aligned. The collective of institutions and the community of research and science has developed since 1945, as the result of a vigorous political debate in the USA as in other western countries. (Kleinman) (Guston & Keniston, 1994; Sarewitz, 1996, 2016) In the USA this resulted in the famous Social Contract for Science in which science was governed by scientists, spending public money without influence or interference by government. Science did rely on its own distinct dynamics in academia, in the different (sub)disciplines, faculties and institutes of universities and in the highly respected and influential learned societies.

Science has thus been established as a 'state in a state', the *Scientific Estate* (Price, 1965) or the *Republic of Science* (Polanyi, 1962b), with its own goals, rules, governance, ethics and (counter)norms (Bourdieu, 2004; Latour, 1987; Merton, 1973; Ziman, 2000) Its culture and politics are until this day largely determined by old ideas originating from the first half of the twentieth century – from the classical philosophy and sociology of science about how science ought to be done. It still has that mythical narrative about 'the scientific method' of the 'hard' sciences, 'pure versus applied', about 'the relation between science and technology' and 'the linear model of innovation' which Vannevar Bush so effectively used to establish the Scientific Estate at the end of WW2. It is still used in the public debates to defend public funding and importantly secure autonomous governance for science and academia (Bush & United States. Office of Scientific Research and Development, 1945; Greenhill, 2000; Kleinman, 1995). Moreover, as I will argue in detail in the following Chapters, it determines to a great deal still our academic culture: how excellence and quality is defined and how choices regarding the research agenda are made but also how it affects diversity and inclusiveness of research and researchers in academia.

1.3   Politics Outside In

In the scientific community and academia there are like in all institutions strong and sometimes opposing ideas at play about how science as a societal force for progress should be organized, positioned and governed in society in relation to need and expectations of the public. How it should be facilitated to be able to show maximal progress and finally and most relevant, how and by whom quality and progress is to be defined. These debates are in some respects quite academic and may sound esoteric but are in fact highly relevant for the daily practice of research. They directly affect questions regarding the effects of internal and external powers in science, for instance regarding internal distribution of credit which involves measures of excellence, academic hierarchies, positions, standing and esteem, allocation of grant money. These politics of science directly and indirectly determine problem choice and thus the growth of knowledge and also the impact of science in a wider societal context. The reader could get the impression from this historical approach and my dominant reference of the seminal works of the second half of previous century, that this mainly is a problem of the past. Make no mistake, unfortunately, that is not the case, as recent empirical work has shown and will be discussed in the following chapters (Fochler & de Rijcke, 2017; Franssen et al., 2018; Hammarfelt & de Rijcke, 2014; Hammarfelt et al., 2017; Kaltenbrunner & de Rijcke, 2016; Müller & de Rijcke, 2017; Rushforth & de Rijcke, 2015; Rushforth et al., 2018; Rushforth & de Rijcke, 2016).

This relates to persistence of classical ideas about the scientific method, truth, value-free science, academic autonomy, neutrality and the insulation of science from external non-scientific values, from politics and from society at large. These problems of engagement and responsibility, versus autonomy and academic freedom are not at all new and have been discussed in the sixties and seventies from different philosophical, sociological and political viewpoints (Rose & Rose, 1969; Ziman, 1996; Ravetz, 1971; Bernal, 1939; Polanyi, 1962b; Habermas, 1970a, b; Toulmin, 1964; Weinberg, 1963).

From 1960 the idea of science as a communitive action, a truly collective and social process with a professional culture organized to produce certified and robust knowledge became established. In particular since Kuhn's *Structure of Scientific*

Revolutions published in 1962, in academia it became acceptable and was recognized indispensable to study the various social aspects of science taking into account, other than strict scientific arguments to be able to explain the growth of knowledge (Kuhn, 1962). These developments originated from ground-breaking work by a few scholars that in a novel way started to perform studies of how science works and how we make and accept knowledge (Hanson, 1958; Toulmin, 1972; Polanyi, 1962a; Ziman, 1968). Their studies deviated from the until then dominant mainly normative philosophical discussion based on the natural sciences on the status of scientific claims, and instead focused on the practice of science and how knowledge and our common beliefs in practice are reached, instead of discussing how science ought to be done. As I will discuss later, despite that this work goes back 40–60 years, since it was mainly performed in the social sciences and humanities faculties, in the faculties of natural sciences, geoscience and biomedical sciences, this still has not been widely noticed. The majority of practitioners and administrators intuitively still go by the Standard Model, a popular image of science that does not correspond with the actual aim and practice of science. The classical ideologized images of science and its poor match with the actual practices of knowledge making, is highly problematic since until this day it determines to a large extend how science is being done in academia. Obviously, a correct self-understanding of science is also of particular importance in debates where proper reflection on the status, the higher purpose and the position of science in society is required (Habermas, 1971).

1.4   The Social Contract of Science Revisited

Despite its own mythical claims of autonomy and of pure investigator-driven research as the highest ideal, science, especially natural science and engineering, but increasingly biomedical research from 1945 on, was driven and had grown immensely by infusion of public money that targeting mainly public issues of health and agriculture. In addition, science around the globe remained heavily connected to the corporate and military sector, also in times of peace after WW2. In the US investments especially the natural sciences were boosted by Eisenhower in reaction to the hysteria in science and education prompted by the first successful manned Soviet spaceflight of Sputnik in 1958 (Greenberg, 1999). In private discussions at that time with his adviser James R. Killian, a former president of MIT, Eisenhower expressed his irritation about scientists who pursued their own interests instead of those of the nation with their work having too little benefit for society and its publics. Interestingly, despite this reprise of Vannevar Bush's agenda of 1945 now provoked by Sputnik and the Cold War, Eisenhower in his famous farewell speech 17 January 1961 expressed his deep worries about the fact that science had been hijacked by the military and the commercial interests of the connected industries.

This boost of the natural sciences in the USA seems in some respect in agreement with C.P. Snow's even more famous, bold and original cultural and philosophical critique of academia and science in his *'The Two Cultures'* (Snow, 1993). He argued in 1959 that academia in the UK held theoretical, pure scholarship of the humanities in much higher esteem than research in the natural sciences with their technical and practical applications. Stefan Collini in his most insightful introduction to the 1993 reprinting of Snow's book, elaborates on its cultural and social background, reception and the brutal dismissal by F.R. Leavis. Collini states the following: *'The 'Leavis-Snow controversy' can obviously be seen a re-enacting of a familiar clash in English cultural history- the Romantic versus the Utilitarian, Coleridge versus Bentham, Arnold versus Huxley and other less celebrated examples(pxxxv.)* Snow, according to Collini, was *'clearly frustrated about the domination of the traditionally educated upper class and was motivated by class resentment'* which places the *'The Two Cultures'* also in a much larger and moral socio-political context than science. In the *Second Look*, indeed Snow confesses that the original title *'Rich and Poor'* would have been better suited for his argument. He ends the book after he discusses major social and economic problems as follows: *'With good fortune, however, we can educate a large proportion of our better minds so that they are not ignorant of imaginative experience, both in arts and science, not ignorant of the endowments of applied science, of the remediable suffering of most of our fellow humans, and of the responsibilities which, once they are seen, cannot be denied'(p100).*

Eight years later, Peter Medawar, a famous observer of science, immunologist and Nobel prize winner, made a similar observation from within his domain of the biomedical sciences. His critique was based on the 'pure versus applied' distinction and he discussed the *'motives which have led people to think* (it) (these different forms of research) *highly important, and above all to make it the basis of an intellectual class-distinction.'* (p120) (Medawar, 1967). *'The two conceptions are, roughly speaking, the romantic and the rational, or the poetic and analytical, the one speaking for imaginative insight and the other for the evidence of senses, one finding in scientific research its own reward, the other calling for a valuation in the currency of practical use' (p10–11). 'The notion of purity has somehow been superimposed upon it (Bacon's distinction) and in a new usage that connotes a conscious and inexplicable sight-righteous disengagement form pressures of necessity and use. The distinction is....between polite and rude learning, between laudable useless and the vulgar applied, the free and intellectually compromised, the poetic and the mundane' (p121–212). While pure science is a genteel and even creditable activity for scientists in universities, applied science, with all its horrid connotations of trade, has no place on the campus' (p126).*

Medawar, who came very early in his life with his parents from Brazil to England and studied at Oxford, came from comparable social backgrounds as Snow, outside the traditional social elites (Collini here cites Trilling pxxxix). Medawar interprets these two distinct conceptions and cultures of science in the larger social cultural Anglo-Saxon context, which was *'terribly English',* he remarks. Both clearly see the unproductive cultural and philosophic tension that even affects the organizational level of the academy. They argue explicitly for a proper balance between science and humanities, but also for a balance of pure and applied within the science disciplines. They see this social and cultural divide and its academic hierarchy as obstructive to optimal societal impact of scientific research and of the academia as an institution. Medawar explicitly discusses criteria of (e)valuation of science, in his opinion being the *'size of ..contribution to that huge, logically articulated structure of ideas' and for humanists 'by different but equally honourable standards, particularly by the contribution it makes, directly or indirectly, to our understanding of human nature and conduct, and human sensibility' (p126).*

Medawar states that 'pure' nor 'applied' are specific criteria for evaluation of research. With the hindsight of 2019, we know, as we shall discuss later, that in the unwritten and written mores they most surely were, and to great extent still are and not only in the UK. Medawar at the same time concludes with a visionary remark that *'The humanist fears that if we abandon the ideal of pure knowledge, knowledge acquired for its own sake, then usefulness becomes the only measure of merit. And that if it does become so, research in the human arts is doomed' (p126).*

This is indeed the major worry from the domain of humanities that Snow and Medawar, despite their complaints could have anticipated based on what had already happened in the politics of science after the 'coup' of the 'hard sciences' and in particular the physicists lead by Vannevar Bush. Interestingly, philosophers in these same years already saw a major problem with the dominance of the worldview and ideas of modernity and the corresponding reductionist positivist Cartesian way of doing science. This 'scientific method' appeared to have proven quite successful first for the technosciences, the 'hard sciences' and later biology and biomedical research but was not appropriate for the social sciences and humanities. Disciplines that studied the social domain of society and human life need the classical pre-modern methods of arguments, reason and rhetoric (Winch, 1958; Toulmin, 1961, 1972). The present-day academic should not forget that SSH were in academia, in sharp contrast to the centuries before, for a large part of the twentieth century, not considered scientific nor serious rational endeavours. It took an extra 8 years before these disciplines were recognized as science and were included after the start in 1950 of the US National Science Foundation. As we will see later in this book, and as Shapin wrote in 2007, Snow was 'not at a funeral of the natural sciences, but at a christening'. *'In the academy and most modern research universities, it is the natural sciences that have the pride of the place and the humanities and social sciences that look on in envy and sometimes resentment.'* (Shapin, 2007).

In the meantime, Project Hindsight, a study on the return of investments in science aiming at military defence ran from 1963 until 1968, that was officially published in 1970. The conclusions were quite shocking for the science establishment. Technology accounted for 91% of the impact, very little was attributable to applied science and nearly nothing to basic science (Sherwin & Isenson, 1967). In these days, the critical comments made by President Johnson at the signing ceremony of Medicare (June 1966), about a lack of clinical impact of publicly funded basic biomedical research ('laboratory research') elicited strong protests from the biomedical research community, which can still be heard in many biomedical institutes.

1.5   The Politics of Scientific Choice

As a logical consequence of these critical views and evaluations of science, the unescapable question of how to deal with 'the complexity of scientific choice' came up. In a series of high-profile papers, published at the very start of the journal *Minerva*, between in 1962–1964. They were written by authorities like Michael Polanyi, Alvin Weinberg, CF Carter and John Maddox and are very reminiscent of the current debates about Incentive and Rewards (Carter, 1963; Maddox, 1964; Polanyi, 1962b; Weinberg, 1963). Stephen Toulmin wrote a review of these papers, discussing the quite different perspectives presented. Up-front he concluded *that 'the questions about selection and priorities, implicit in all discussions of science policy are both difficult and inescapable'* (Toulmin, 1964). The problem, he writes is there for both less developed and industrialised (developed) countries, but they are of course very different for them (p333). The difficulty is that we know too little about the consequences and long-term impact of in particular fundamental research since we know too little about the course that both science and society will run and which problems will emerge. Toulmin suggests that we therefore should systematically study sociological, economic and organisational questions involved in the interactions between science and society. In addition, he concludes that we need to understand the issues at play in the formulation and administration of a science policy and 'remove any fog due to ambiguities, cross-purposes or hidden assumptions'. Polanyi is well known for his advocacy for the autonomous and self-governed 'Republic of Science'. Its higher aim being to reveal *'a hidden reality for the sake of intellectual satisfaction'*. He argues strongly for the scientific community and its internal structures to decide on scientific choice. *"Guiding the progress of science into socially beneficent channels'* is *'nonsensical'* and *'guiding scientific research towards a purpose other than its own, will deflect the advancement of science'* (cited by Toulmin). Maddox agrees with him, pointing out that it takes debates between academics (intellectual confrontations and open discussions) to decide on research priorities, which he says will also have to be done for the technological applied sciences. This is all well, hard to do, even within a given branch or subdiscipline, but Weinberg taking the problem to a higher organizational level, is more interests in choices *'which pit different fields against each other, for instance molecular biology, high energy physics and behavioural sciences'*. Their potential impact and relevance in science and society is incommensurable. He proposes and elaborates on three criteria's of merit: technological, scientific and social. For massive public support at least two should be highly rated. Social merit is to be decided on external arguments (politics and values) about issues like *'health, food production, defence and prestige'*. He, being a physicists, offers some judgement: *'molecular biology has all three merits, but high energy physics is somewhat overrated,... space-research is only masquerading as science, but if it is more on prestige (first man on the moon) or for military impact we should say so'*. Carter comes from a very utilitarian economic perspective and regarding pure research he believes that *'any nation is at liberty to undertake pure research beyond its justification by its ultimate application'*. There is of course no one science policy, says Toulmin, and he points to the many science policy choices that continuously have to be made, in science and governments, as the way these obvious different perspectives, play out in reality. Because of the plurality of problems, in science and society they will be plurality of criteria and merits that are relevant in the many different contexts in which *political* choices regarding science policy have to be made. He also points out the problematic use of 'the scientists' and 'the scientific community', the lack of democracy of these communities with its 'age-and-status structure' of a *gerontocracy* which impedes assessing *'the* scientific opinion'. There are many interactions and contact points of science with government and these involve many different scientists who will apply 'their minds to a different group of problems and the needs of each partnership will impose their own pattern of research priorities and criteria of choice'. Regards this debate, Toulmin distinguishes also four distinct types of research, from (1) pure natural science to (2) speculative technology, to (3) applied product- oriented and (4) problem-oriented research aimed at solving a particular practical problem that has different stakeholders in science and society. He continues this paragraph with an insightful statement reminiscent of John Dewey's pragmatism that at that time already was nearly forgotten: *"The urgent question to-day is, how the republic of science is to be integrated not only into the broader academic confederation, but into the whole community of citizens. For it is on the answer to this question that our broader criteria of scientific choice ultimately depend'.*

This thinking was propagated before in the first wave of Science for Society in the UK in 1935 by the so-called 'scientific humanists' including J.D. Bernal, Frederick Soddy and colleagues with their book *'The Frustration of Science'* and the founding in 1938 of a new division of the British Association for the Advancement of Science for the social guidance to the progress of science. Followed by the Royal Society's initiative put to the universities in 1945 for 'The Balanced Development of Science in the United Kingdom. All 'at best pointless' in the opinion of Polanyi. For Polanyi opening up science and research to politics and publics and being held responsible for the adverse effects of its research, was an absolute 'no go', which was based on his traumatic experiences in the less open and less democratic societies he had fled from (Guston, 2012). He was thus happy to conclude in 1962 that 'this movement (by Bernal and colleagues) has virtually petered out'. He asks the for him rhetoric question *'Have not even the socialist parties throughout Europe endorsed now the usefulness of the market?"* We will see that maybe they did not in 1962, but they really did from 1980 on. His own *'Society for Freedom in Science'*, however, established in reaction to Bernal et al., after its start in 1944 was also very short lived (Society for Freedom in Science, Nature, July 8, 1944).

1.6   Conclusion

As of this writing, the relations and interactions between science and society and the issues of problem choice for the setting of the science agenda, obviously are still topics of hot debate. They touch upon many crucial aspects of the practice of science, but also on the dangers of the possibility of abuse of science via the immense powers of multinationals in our deregulated neoliberal economies. We have to keep in mind the threat to free scholarship and research in many countries where democracy itself is under threat. Before I discuss the more recent developments in light of these images of science in Chaps. 5, 6 and 7, I will analyse in more detail which images of science are involved, what their status is and where they originate from (Chap. 2). Then I will discuss how they determined and distorted our views, attitudes, policies and the organization and potential of science and its interactions with stakeholders in society (Chap. 3).

References

Bachelet, M. (2019). Data on child deaths are a call for justice. *Nature, 574*(7778), 297. https://doi.org/10.1038/d41586-019-03058-6

Beck, U. (1992). *Risk society towards a new modernity*. Sage.

Bernal, J. D. (1939). *The social function of science*. G. Routledge & Sons ltd..

Bourdieu, P. (2004). *Science of science and reflexivity*. Polity.

Boyd, I. L. (2019). Scientists and politics? *Science, 366*(6463), 281–281. https://doi.org/10.1126/science.aaz7996

Bush, V., & United States. Office of Scientific Research and Development. (1945). *Science, the endless frontier a report to the President* [text] (pp. 1 online resource (ix, 184 pages illustrations)). Retrieved from HathiTrust Digital Library. Freely available. http://catalog.hathitrust.org/Record/001474927

Carter, C. F. (1963). The distribution of scientific effort. *Minerva, 1*(2), 172–181. https://doi.org/10.1007/BF01096249

Cohen, H. F. (2010). *How modern science came into the world: Four civilizations, one 17th-century breakthrough*. Amsterdam University Press.

Coyle, D. (2019). When capitalisms collide. *Nature, 574*(7778), 322–324. https://doi.org/10.1038/d41586-019-03047-9

Dewey, J. (1948). *Reconstruction in philosophy* (enlarged ed.). Beacon Press.

Fochler, M., & de Rijcke, S. (2017). Implicated in the Indicator game? An experimental debate. *Engaging Science, Technology, and Society, 3*, 20. https://doi.org/10.17351/ests2017.108

Franssen, T., Scholten, W., Hessels, L. K., & de Rijcke, S. (2018). The drawbacks of project funding for epistemic innovation: Comparing institutional affordances and constraints of different types of research funding. *Minerva, 56*(1), 11–33. https://doi.org/10.1007/s11024-017-9338-9

Greenberg, D. S. (1999). *The politics of pure science*. University of Chicago Press.

Greenhill, K. M. (2000). Skirmishes on the "endless frontier": Reexamining the role of Vannevar bush as progenitor of U.S. science and technology policy. *Polity, 32*(4), 633–641. https://doi.org/10.2307/3235296

Guston, D. H. (2012). The pumpkin or the Tiger? Michael Polanyi, Frederick Soddy, and anticipating emerging technologies. *Minerva, 50*(3), 363–379. https://doi.org/10.1007/s11024-012-9204-8

Guston, D. H., & Keniston, K. (1994). *The fragile contract : University science and the federal government*. MIT Press.

Habermas, J. (1970a). *Toward a rational society*. Heinemann Educational Books.

Habermas, J. R. (1970b). *Towards a rational society: Student protest, science and politics*. Beacon Press.

Habermas, J. (1971). *Knowledge and human interests*. Beacon Press.

Hammarfelt, B., & de Rijcke, S. (2014). Accountability in context: Effects of research evaluation systems on publication practices, disciplinary norms, and individual working routines in the faculty of arts at Uppsala University. *Research Evaluation, 24*(1), 63–77. https://doi.org/10.1093/reseval/rvu029

Hammarfelt, B., de Rijcke, S., & Wouters, P. (2017). From eminent men to excellent universities: University rankings as calculative devices. *Minerva, 55*(4), 391–411. https://doi.org/10.1007/s11024-017-9329-x

Hanson, N. R. (1958). *Patterns of discovery an inquiry into the conceptual foundations of science*. Cambridge University Press.

Kaltenbrunner, W., & de Rijcke, S. (2016). Quantifying 'output' for evaluation: Administrative knowledge politics and changing epistemic cultures in Dutch law faculties. *Science and Public Policy, 44*(2), 284–293. https://doi.org/10.1093/scipol/scw064

Kleinman, D. L. (1995). *Politics on the endless frontier : Postwar research policy in the United States*. Durham Duke University Press.

Kuhn, T. (1962). *The structure of scientific revolutions*. 2nd edn, enlarged 1970. University of Chicago Press.

Latour, B. (1987). *Science in action : How to follow scientists and engineers through society*. Harvard University Press.

Maddox, J. (1964). Choice and the scientific community. *Minerva, 2*(2), 141–159. https://doi.org/10.1007/BF01096591

Medawar, P. S. (1967). *The art of the soluble*. Methuen.

Merton, R. K. (1973). *The sociology of science: Theoretical and empirical investigations*. University of Chicago Press.

Müller, R., & de Rijcke, S. (2017). Thinking with indicators. Exploring the epistemic impacts of academic performance indicators in the life sciences. *Research Evaluation, 26*(3), 157–168.

Nowotny, H. (2016). *The cunning of uncertainty*. Polity.

Polanyi, M. (1962a). *Personal knowledge ; towards a post-critical philosophy*. Harper Torch Books.

Polanyi, M. (1962b). The republic of science. *Minerva, 1*(1), 54–73. https://doi.org/10.1007/BF01101453

Price, D. K. (1965). *The scientific estate*. Belknap Press of Harvard University Press.

Ravetz, J. R. (1971). *Scientific knowledge and its social problems*. Clarendon Press.

Rose, H., & Rose, S. P. R. (1969). *Science and society*. Allen Lane.

Rushforth, A., & de Rijcke, S. (2015). Accounting for impact? The journal impact factor and the making of biomedical research in the Netherlands. *Minerva, 53*(2), 117–139. https://doi.org/10.1007/s11024-015-9274-5

Rushforth, A. D., & de Rijcke, S. (2016). Quality monitoring in transition: The challenge of evaluating translational research programs in academic biomedicine. *Science and Public Policy*. https://doi.org/10.1093/scipol/scw078

Rushforth, A., Franssen, T., & de Rijcke, S. (2018). Portfolios of worth: Capitalizing on basic and clinical problems in biomedical research groups. *Science, Technology, & Human Values, 44*(2), 209–236. https://doi.org/10.1177/0162243918786431

Sarewitz, D. R. (1996). *Frontiers of illusion : Science, technology, and the politics of progress*. Temple University Press.

Sarewitz, D. R. (2016). Saving science. *The New Atlantis, 49*, 4–40.

Shapin, S. (1996). *The scientific revolution*. University of Chicago Press.

Shapin, S. (2007). Science in the modern world. In O. A. E. Hackett, M. Lynch, & J. Wajcman (Eds.), *The handbook of science and technology studies* (3rd ed., pp. 433–448). MIT Press.

Sherwin, C. W., & Isenson, R. S. (1967). Project hindsight. *Science, 156*(3782), 1571–1577. https://doi.org/10.1126/science.156.3782.1571

Snow, C. P. (1993). *The two cultures*. Cambridge Unversity Press.

Toulmin, S. (1961). *Foresight and understanding; an enquiry into the aims of science*. Indiana University Press.

Toulmin, S. (1964). The complexity of scientific choice: A stocktaking. *Minerva, 2*(3), 343–359. https://doi.org/10.1007/BF01097322

Toulmin, S. (1972). *Human understanding*. Clarendon Press.

Weinberg, A. M. (1963). Criteria for scientific choice. *Minerva, 1*(2), 159–171. https://doi.org/10.1007/BF01096248

Winch, P. (1958). *The idea of a social science and its relation to philosophy*. Routledge and Kegan Paul Humanities Press.

Ziman, J. (1968). *Public knowledge : An essay concerning the social dimension of science*. Cambridge University Press.

Ziman, J. (1978). *Reliable knowledge : An exploration of the grounds for belief in science*. Cambridge University Press.

Ziman, J. (1996). Is science losing its objectivity? *Nature, 382*(6594), 751–754. https://doi.org/10.1038/382751a0

Ziman, J. M. (2000). *Real science : What it is, and what it means*. Cambridge University Press.

Chapter 2
Images of Science: A Reality Check

Abstract It will be argued that the dominant form of current academic science is based on ideas and concepts about science and research that date back to philosophy and sociology that was developed since the 1930s. It will be discussed how this philosophy and sociology of science has informed the ideas, myths and ideology about science held by the scientific community and still determines the popular view of science. It is even more amazing when we realize that these ideas are philosophically and sociologically untenable and since the 1970s were declared obsolete by major scholars in these same disciplines. To demonstrate this, I delve deep to discuss the distinct stages that scholars in philosophy, sociology and history of science since 1945 to 2000 have gone through to leave the analytical-positivistic philosophy behind. I will be focusing on developments of their thinking about major topics such as: how scientific knowledge is produced, the scientific method; the status of scientific knowledge and the development of our ideas about 'truth' and the relation of our claims to reality. It will appear that the positivistic ideas about science producing absolute truth, about 'the unique scientific method', its formal logical approach and its timeless foundation as a guarantee for our value-free, objective knowledge were not untenable. This is to show how thoroughly the myth has been demystified in philosophy and sociology of science. You think after these fifty pages I am kicking a dead horse? Not at all! This scientific demystification has unfortunately still not reached active scientists. In fact, the popular image of science and research is still largely based on a that Legend. This is not without consequence as will be shown in Chap. 3. These images of science have shaped and in fact distorted the organisational structures of academia and the interaction between its institutes and disciplines. It also affects the relationship of science with its stakeholders in society, its funders, the many publics private and public, and policy makers in government. In short, it determines to a large degree the growth of knowledge with major effects on society.

F. Miedema, *Open Science: the Very Idea*,
https://doi.org/10.1007/978-94-024-2115-6_2

In this chapter, but throughout the book, I will present a narrative in which I will take my own intellectual and scientific journey from 1971 as a chemistry student who did a minor in the philosophy of science in academic year 1975–1976. Since then, I followed the classical career path of a professional biochemist/immunologist, as PhD student, post-doc, group leader, department head, director of a small research institute, to finally become dean and board member of a large University Medical Centre. Going through this professional sequence, I kept a persistent and ever stronger interest in the science of science. It is from the perspective of a true understanding of the practice of science in its various aspects that I will use specific authors a lot, but others much less or even neglect work of many scholars that to specialists in the different fields are considered important but are of little or no relevance for the daily practice of active researchers and most other actors in the field.

2.1   Part 1. Images of Science, a Reality Check

'The empirical basis of objective science has thus nothing 'absolute' about it. Science does not rest upon solid bedrock. The bold structure of its theories rises, as it were, above a swamp. It is like a building erected on piles. The piles are driven down from above into the swamp, but not down any natural or 'given' base; and if we stop driving the piles deeper, it is not because we have reached firm ground. We simply stop when we are satisfied that the piles are firm enough to carry the structure, at least for the time being'. (p109) (Popper, 1959)

Introduction

Unlike most natural scientists writing about science that are not philosophers or amateur philosophers like me, I am convinced that I need to discuss the origins of the philosophical ideas and concepts that are the basis of the dominant image of modern science that in 1981 still was *'the widespread popular conception of science'* (p2) according to Ian Hacking in his influential book *Representing and Intervening.* (Hacking, 1983) I experienced time and again during my professional career that it are these obsolete and incongruous ideas about science and research that even now determine and distort to a large extent our views, attitudes, policies and politics, discourse, professional and collegial interactions in academia. I fully realize that the analysis that follows, to readers with less than average knowledge of the history of the philosophy of science, may feel as a much too deep dive. Understandably, they will wonder whether they need to know all that. The story of analytical philosophy and logical positivism and how it has impregnated our image of science, is essential for my argument to understand the origins and persistence of the problems of science and academia. One can without a problem skip, the whole or Part 2 of this chapter and only take note of the conclusions of Part 1. For a more general quick read, I refer to Chap. 3 of my *Science 3.0* (Miedema, 2012) or the very nice paper by (Pinch, 2001) or Shapin's *Science and the Modern World.* (Shapin, 2007).

The Frontstage and Backstage Paradox

The popular image of science, mainly of the natural and biomedical sciences is sometimes called the Standard Model. It is the well-known narrative of 'the scientific method' and 'the vocational noble' scientists discovering nature and truths'. It is based on a blend of normative philosophy, mainly of epistemology designated the 'Legend' and normative sociology, both were developed in the first half of the twentieth century. This romanticised image is still widely used 'on stage' in the media, in public debates not only when science is besieged or if scientists feel besieged or fear budget cuts. Paradoxically, contrary to this 'frontstage' image, most scientists, 'backstage' in their training and daily professional life are somehow aware that there is no unique method, no formal logic which guides scientists to the truth. In contrast, when being introduced to the daily research practice, they are trained to use a set of instrumentalist principles and methodologies how to make reliable knowledge. Most of these are practical principles referring to techniques, producing and reading texts being journal articles or books, how to set up experiments or investigations, about interpreting and discussing experimental results, the requirement of reproducibility, and thus how to conclude what is to be believed or if you will, is 'true'. These are being passed on to new generations of researchers while they are doing their first rotations in laboratories and departments as master students or PhDs. Of course, there are courses on methods in the field of research -for instance in my case as a BSc chemistry and MSc biochemistry/immunology student since 1971 chemistry, biochemistry, immunology, bacteriology, virology, molecular biology-, and on methodologies like epidemiology, statistics, bioinformatics, spectroscopy, mass spectrometry, NMR, fMRI, genetics. Students are introduced to the state of the art of the discipline with its most novel technical developments and findings. In the natural and biomedical sciences introduction is done almost without reference to history, the pathways that led to that state of the art in the field.

We, as natural scientists do not worry too much about a formal timeless foundation on which we build our investigations, experiments, claims and conclusions. The most important thing you learn is that your claims must hold, that is, can be successfully used by others inside or outside the laboratory or department. Those exceptional scientists who started to think and write about science did not spent too many words on the philosophy and sociology of science. In the natural and life sciences one can become a tenured professor without ever having to read or having read Popper, Merton or Kuhn although most of them want us to believe they once did. There is slightly more interest in the history of the sciences, which mostly are romanticized narratives about the classical gems with an even more classical linear narratives explaining how we arrived were we are now, with a lot of attention for the top scientists, the geniuses in the field. These histories until the 1970s were almost all written from the perspective of the Standard Model. The most famous and widely read exception still is James Watson's *The Double Helix* published in 1968 which for that reason had a very critical reception that still is of great interest to our understanding of images of science and scientists for which I will return to below. (Watson, 1968)

The Standard Model and the Legend

Still the best-known image and narrative of science, of how inquiry and research is being done, I am afraid, is an idealized picture that has in the literature been designated the Standard Model sometimes also called the popular view. The Standard Model is an interesting composite. Its image is built on the one hand on the classical theory about scientific investigation, its unique method, the status of its knowledge claims and the belief system associated with it. This image coming from the philosophy of science has been designated 'the Legend'. Indeed, until this day, implicitly but also explicitly very much of the Standard Model echoes the ideas of what used to be the dominant philosophy and sociology of science until the 1960s.

These ideas about the theories and statements of science and the unique formal status of its knowledge claims, have been developed in the philosophy of science in the first half of the twentieth century. This originated from the seventeenth century Cartesian rationality of *Modernity* which takes its name after Descartes. There are some influences from early positivists like Comte but its form is mainly determined at the beginning of the twentieth century when it became admixed with elements of the logical positivistic tradition of the Vienna Circle, the analytical philosophy of science and the works of Popper. Descartes assumed a formal mathematical method that would be grounded on a set of timeless universal principles, an objective foundation and even unique 'God-given endowments to the human mind' were invoked (Descartes, 1968). This would be the general solution to the problem of the logical formal relation between the observed and the observer. The positivists and Popper, however rejected this timeless and objective *'God's eye perspective'* or 'Archimedean point' as metaphysics, non-empirical and thus per definition as unscientific. To deal with the problem of objectivity- how can we objectively know without our own cultural biases and hidden personal values – an independent analytical foundation for the logical relations between theoretical statements and statements about observed entities and facts was postulated. The prominent members of the Vienna Cycle (Wieners Kreis) in the years before the second World War sought refuge in the USA and there started departments of philosophy in different universities there. In these departments with their approach to philosophy of science, in the analytical, empirical or logical- positivistic tradition, they made school. As a consequence, this philosophy was dominant for a long time around the globe. For a highly readable and informative history of the Vienna Circle see David Edmonds, 'The murder of Professor Schlick'. Popper, was peripheral to the Vienna Circle, spent the years of the war in New Zealand and returned to London after the war. He had realized already that observational statements are theory-laden and eventually concluded that there is no *'given'* foundation, no formal set of principles to build on. He wrote, *'we are drilling piles in quicks and until they stand, and we can build on them for the time being at least'*. We believe and accept or reject theories after serious experimental testing and scientific debate about the evidence he said (Popper, 1959). This Popperian fallibility reminds of Charles Sanders Peirce' early works on how and why we believe, published in the last decades of the nineteenth century.

The Standard Model thus explicitly, via the Legend largely follows the hypothetico-deductive cycle of proposing hypotheses and its derived statements, experimental testing of these statements, with the result of falsification or support or partly support from the observed evidence. This results in acceptance ('belief') or requires improvement and a new cycle of testing. From lower-level observational statements and laws, higher level ever more general laws are deduced which ideally conjecture universal and timeless truth as the most prominent results of scientific inquiry. The reductionist method it proposes is empirical, formal, logical and thus importantly a guarantee for objectivity, because it separates values from positive facts, scientific from non-scientific statements (Nagel, 1961; Hacking, 1983) The strict Cartesian dualism between observer and observed, between fact and value and between analytic and synthetic makes science per definition reliable, because its products are objective, value free and thus trustworthy. It was for a long time self-evident that this 'scientific method', with its rigor and potential for prediction and control building on the ideal of Euclidian mathematics, was the cause of the over-whelming theoretical and practical technical successes of the natural sciences. It so happened that positivism and Popperian demarcation of falsification between scientific and non-scientific knowledge became dominant.

It moreover, was generally believed to be the critical difference between the natural 'hard' sciences and the 'soft' social sciences and humanities. This demarcation is about methods of investigation, but also about its products, its theories and laws which can be tested and in the hard sciences preferably were expressed formally thus mathematically and were held to be universally true. If investigation was performed in that tradition and thus modelled after the natural sciences, especially after physics, it would be recognized as science. Given its main philosophical sources, the type of research of the Standard Model aims for the ideal of timeless universality, wants to produce general laws, formal basic knowledge using reductionist methods to contribute to the body of knowledge. It is historically mostly confined to the classical academic disciplines and operates in an international global perspective. It aims for value-free research and neutrality, is in principle against interference from whatever powers outside academia or even from within academia outside the own discipline.

Based on its own criteria for what is considered to be science, research done in this way always was, and to a large degree still is the highest in rank within academia compared to the social sciences and the humanities (SSH). SSH until 1958 not in the least for this reason was not regarded serious science or research and for instance not a discipline in the National Science Foundation in the USA. As I will argue in later chapters in more detail, still in the third decade of the twenty-first century, within virtually every discipline and faculty, there is a visible gradient of research esteem according to the degree of the use of formal quantitative methods that employ or at least imitate the methods of the Legend and thus of the natural sciences.

The Mertonian Social Order

The Standard Model is a composite of the Legend of the scientific method described above, but in addition, explicitly builds on the classical sociological image of science which has originally been developed by the famous American sociologist Robert Merton and his students between 1930 and 1970. (Merton, 1973) In this image of science, it is a human activity different from all other human activities in that scientists are altruistically looking for the truth. This is, according to Mertonian sociology, done in a open community, chatacterized by sceptical debates about each other's work in order to get to the best knowledge. Knowledge is considered or at least aims to be universal and not bound or restricted to time and place. Importantly, the scientists are fair in discussing the works of their peers and are honest or at least strive for honesty. They are not in it for their own personal or intellectual interests. They publish their results for their peers to judge and to be used for further research. Their findings are thus expected to be made freely available and in all respects are considered common good. They can through the workings of the incentive and reward system, commissioned by the scientific community, get credit for their work, which is required to advance their careers and gain in reputation and standing in their respective field of research. Reputation is gained for instance by so-called 'priority', being the first to discover and report facts, theories and novel methods, and contributions that by peers are considered relevant and original. In this vision there is fierce competition and consequently to it stratification. There are elites in every discipline, which in the Mertonian social order is however not felt to be (too) problematic, but is considered instrumental for the functioning of the enterprise and thus reflects the natural order, a logical consequence of the type of activity the community is engaged in (Ben-David & Sullivan, 1975). Merton in 1968 did however already point out several unwanted effects of stratification inherent to the reward system (Merton, 1968). Although all researchers are in principle regarded equal, elitism is acknowledged but thought to be functional. Merton coined the term *Matthew Effect* for the famous, or more recently considered, infamous mechanism of accumulative advantage that elites in the system have. These advantages concern influence, authority and professional power which gets converted in material advantages like, research facilities, grant support and access to the most prominent academic functions and positions. If you read the paper more then 50 years later, you are struck by the normative and outright naïve and idealistic wordings by which Merton describes his expectations how the top scientists will deal with or even counteract any perverse effects of the *Matthew Effect* if it would ever become 'an idol of authority'. He has amazing faith in top scientists because of their unusual characters and high standards of integrity (Merton, 1968). In adhering to the norms, and so producing results and publications, scientists are recognized as good citizens by their peers and members of the community and accepted and respected as members of the scientific enterprise. Moreover, by keeping up this academic social culture, science, it is believed is trusted and earns respect from the public and government as a reliable institute in society. In the Mertonian view, science is a closed social system within society that decides itself who is excellent and who is not, who gets the credits, the jobs and the grant money. This implicates that the growth of knowledge in this view is an internal affair. Science is a value-free, neutral, activity where autonomous individuals disinterestedly pursue their inquiries in the context of a social system governed by its own unique internal scientific criteria and norms.

Dispatches from the Trenches

I realized the problems the popular image of science, held by the science community and the public and started to study it, in the early 1980s during the start of my scientific career as a researcher on the pathogenesis of aids and HIV infection. That was in a truly unique setting in which my group, or as we say in our field 'my lab', worked on HIV/aids in Amsterdam in a cohort study of men who have sex with men (MSM) and IV drug users. In these Amsterdam Cohort Studies it had been clear from the start in 1985 that to understand the problem of aids and HIV infection, a truly multidisciplinary approach was needed. My colleagues came from the social and behavioural sciences, medical anthropology, epidemiology of infectious diseases, bioinformatics, internal medicine, pathology, pre-clinical and medical virology. Next to this array of scientific disciplines we interacted proactively with the participants of the cohorts, mainly homosexual men. Listening to their concerns, their problems and immediate needs but also to keep them informed about the work we did using their blood samples and the epidemiological and behavioural information they provided in the questionnaires. The work was done the Municipal Health Centre, AMC and my group was working on viro-immunology in the Central Laboratory of the Blood Transfusion Service (CLB, now Sanquin). At my institute with respect to aids, research was done in the wider context of the safety of blood supply which was at that time of the highest daily concern. This bloodbank context involved cellular and protein chemistry, virology and technical issues of manufacturing of biologicals, but also sociology, economics and ethics of blood donation and screening of donors.

I read Latour's *Science in Action* in 1987, as a young principal investigator working on HIV/AIDS already getting deep into international science (Latour, 1987). The researchers that Latour followed in the lab and outside the lab talking to the different stakeholders, on their travels abroad were pretty busy. All of it was familiar to me. Only years later I discovered a major early source of Latour, Bourdieu who applied his theory of the 'field' to academia with its concepts of habitus, socialization, the power struggle, stratification and elitism (Bourdieu, 1975, 2004). Few biochemists or natural scientists in their scarce time do read such scholarly studies about themselves, despite the insightful analysis of the familiar academic microcosm which we virtually on a daily basis were deeply involved. It made me aware of quality and credibility, the standing of the different sciences and institutions, about competition and power games, reputation, getting credit, about the moral values and the

(continued)

personal motivations involved in science, that implicitly and explicitly could be observed in daily verbal and non-verbal interactions.

After spending 35 years in that multidisciplinary environment in a highly competitive national and international world of science it was obvious that scientists from different fields and disciplines see the world differently and speak different languages. These are, however, minor issues compared to the much more serious and also widely held misconceptions and prejudices about research and inquiry, about the different academic disciplines and what the true aims of science are. These appeared to be mostly based on obsolete ideas derived from the classical philosophy and sociology of science.

This would not be a problem,

if it would not have adverse effects at the national or institutional level, for instance on agenda setting and the growth of knowledge

if this would not cause major science waste and production of much poorly performed and useless research

if this would not be the cause of major obstacles for translation of research to societal impact for those in the real world who need solutions and relief badly.

Unfortunately, daily experiences in the community of science already over a very long time show differently. It did and until this day does cause various serious problems that affect science and inquiry at many levels and affects its potential to impact society. It is because of this that I will in more depth discuss the popular images of science, their origins and problems and how they affect the practice of science. After that I will in this Chapter discuss the philosophy and sociology that forms the foundation for these popular images and discuss how these ideological and normative concepts, with their respective famous dualisms have in the past 40 years been shown by philosophers, sociologist and historians of science to be scientifically untenable.

The Mythical Image of Science

The Standard Model thus is an image of science that is a composite of two narratives, based on a philosophical and a sociological theory established in the first half of the twentieth century. First there is a powerful ideal, derived from philosophies based on the natural sciences with an implicit positive image of scientist's intentions and social interactions, in which the unique relation between theories and its knowledge claims with reality stand out. Next there is the sociological image of a community of vocational altruistic investigators who in daily practice go through daily struggles and hard labour to discover the secrets of nature and come to a set of unifying ideas about the world. The Standard Model does not present a consistent idea of science because these two components synergize but fail to merge into an overall *theory* of science that explains **how science really works and how that relates to its reliability, success and credibility.** It is exactly because of this hybrid, with these two complementary faces, the Standard Model as an image and a general narrative about science has worked well for science in its interaction with the outside world in the past.

Obviously, it has had its value and advantages, but it is I will argue, also since long the root cause of the most urgent problems in the relation between science, government and society, and at the lower level in academia, between scientists and between scientists and their publics. Both aspects of the popular image or science described above do not resonate much with active researchers. The way we have made and make knowledge that works and leads to successful follow-up investigations and subsequent growth of knowledge as well as successful interventions in the real world, the practice of science in the natural sciences including physics, is fundamentally different from what the Legend holds on philosophical grounds to be the unique scientific method to arrive at true, believes, statements and insights. Active researchers in the different fields and disciplines do not pay too much attention to the rules of engagement of the Legend as far it concerns the celebrated scientific method. They don't need to. In addition, with respect to the Mertonian norms, there are written codes of conduct and written and tacit mores, that researchers intuitively and indirectly are aware of it. As soon a sociologists started to actually take a look at the practice of science, they couldn't help themselves seeing major and general aspects of behaviour and mores of active researchers not in agreement with the Mertonian ideal. This was observed at the individual level, but also at the institutional level. This has in the past 10 years increasingly drawn attention within the scientific community and lately this was discussed in the media and public debates as well (Chap. 3).

The Standard Model: A Reality Check

I will discuss the criticisms that have started to develop mainly since 1960 regarding the philosophical theory as well as on the sociological theory that formed the main pillars of the Standard Model. These criticisms are based on research in philosophy, sociology but also history of science. We will see that both components of the model have been shown to be normative in nature, not reflecting nor impacting much the practice of the sciences.

Possessed by the Normative, Demeaning the Descriptive

Philosophers have long made a mummy of science. When they finally unwrapped the cadaver and saw the remnants of an historical process of becoming and discovering, they created for themselves a crisis of rationality. That happened around 1960. It was a crisis because it upset our old tradition of thinking that scientific knowledge is the crowning achievement of human reason. Sceptics have always challenged ...but now they took ammunition from the details of history (p1) (Hacking, 1983)

As described in the previous section, until 1960 the dominant philosophy of science was based on concepts and ideas developed in the empiricist and logical positivistic tradition very much inspired and lead by the way of thinking of analytical philosophy. It is totally devoid of historical perspective and did not at all take into account the diverse research practices, the way research was being done and thus how in the laboratory we actually produce knowledge and decide what to belief. Even in recent times, members of the scientific community, when being asked, still belief in the ideals and norms of the Standard Model. Although deep inside they know that at the organizational and at the personal level science has never functioned according to these rules and norms, as sociological and historical researchers have demonstrated in the past 40 years. (Hanson, 1958; Toulmin, 1972; Kuhn, 1962; Ravetz, 1971; Ziman, 1968, 1978; Latour, 1987; Latour & Woolgar, 1979; Mitroff, 1974; Shapin, 1982). Furthermore, although the foundations and the logic of the scientific method were questioned already since the 1930s, in several disciplines, −biology, medicine, economics, including the social sciences- subdisciplines and research fields emerged that copied the formal quantitative methods and style of research of the 'hard' sciences. They have a craving for the type of science that never was which is also called 'physics envy'. Toulmin for the field of economics describes this development in a chapter under the title 'Economics and the Physics that never was'. (Toulmin, 2001).

As we already saw, which in this light is truly remarkable, the ideas, or as some say images, of science in these philosophies were by most scientists not only taken for granted but also somehow believed to be descriptive. One wonders why the science community and the public did (does) go along so well with the Legend. Was it despite the fact, or is it because it is normative and ideal, and not in any sense related to how science was done in practice? Do we all still very much want to believe and hope that science is really different from all other human activity and do we like to deem scientists as virtuous and pious as the high priests and cardinals that never where. Even when confronted with flagrant deviations, when the Legend is in doubt *'there is often a significant shift in perspective. The image is no longer seen as descriptive but normative. Despite this shift, a connection with description usually remains. The problematic work is a deviation from the proper course of scientific activity, a course taken to be exemplified in the overwhelming majority of scientific investigation.'* (Barker & Kitcher, 2013).

In his *'Human Understanding'* published in 1972, but also in his illuminating earlier and later work, Toulmin was one of the first to see this separation of the practice of knowledge from its theory as the major problem in our theories about science and research and thus of human understanding. Early in his career in Oxford he says: *'This was seen as being quite separate and independent and so a concern of different intellectual professions. At these times, natural scientists kept their eyes outwards, so as to avoid becoming entangled in philosophical word-splitting'p1.* But he continues *'There are in fact good reasons, both historical and substantial, for our establishing links between the scientific extension of our knowledge and its reflective analysis and reconsidering our picture of ourselves as knowers in the light of recent extensions to the actual content of our knowledge.' (p2).* On that same second page he already anticipated anxiety, uncertainty and scepticism, but he reassured the reader that *'a realistic appraisal of human understanding has often been an instrument for its systematic improvement'.* (Toulmin, 1972).

Toulmin could have known better, his early work in the 1950s took a different position on rationality and reasoning from the then mainstream philosophy. His ideas about the philosophy of science were inspired and in effect went through a reality check when he was being exposed in the war to real physics research and the actions of researchers in the lab. After the war he returned to study with Ludwig Wittgenstein who in those days had reconsidered the formal approach in analytical philosophy. Toulmin took up the historical approach to studying science in a natural way blended with philosophy and sociology. In this 'historical turn' he was a front runner and was therefore side-lined and largely neglected for three decades by mainstream philosophy (Toulmin, 2001), which as Shapin wrote, still did hurt after 40 years (Shapin, 2002). Interestingly, in line with my own experience as a student from 1975 on, those who in those days started to study the philosophy of science, somewhere in their career of an experimental natural scientist, gradually realized that the philosophy and sociology did not relate to practice of the natural science.

Introduction into Philosophy of Science

After obtaining a bachelor's degree in chemistry from the University of Groningen, I spent the academic year 1975–1976 studying philosophy of science. In my master study it was a minor with a major in biochemistry. This was inspired by my older brother who studied in the same period history and philosophy of education and philosophy in Groningen. Had my older brother chosen to study theoretical physics instead of pedagogy and philosophy, the course of my intellectual and personal life would most likely have been very different. Because I was completely ignorant, I had to study in the spring of 1975 as introduction the first 300 pages of Ernest Nagel's *The Structure of Science: Problems in the Logic of Scientific Explanation* (Nagel, 1961) in combination with Toulmin's more idiosyncratic *Philosophy of Science* (Toulmin, 1953). This was meant to be a high-speed introduction to be able to study Kuhn's *'Stucture'* and Poppers *'Logic'* followed by an intensive winter-course on the seminal book *'Criticism and the Growth of Knowledge'*, edited by Lakatos and Musgrave (1970). I found the image and discussion of science in Toulmin's book logical and his metaphor of maps for theories plausible. I

(continued)

recognized a lot of common sense in the description of instrumentalism by
Nagel (1961, p129–140). Instrumentalism was down-played very much compared to the overwhelming emphasis on the natural sciences, mathematics,
geometry and physics and its empiricism and logical axiomatic systems of
positivism. For me, despite my chemistry bachelors with introductions in
math, chemistry, biophysics but even some quantum physics, it was simply
too much. Until very recently I labelled Nagel as a diehard logical positivist.
I however should have paid more attention to the introduction of his classical
book. Nagel clearly shows his preference for pragmatism in the Peircean style
which is a plain critique of the empiricist-positivist philosophy of the
'Legend'. I also could have paidd attention to his references to C.S. Peirce,
Frank Ramsey and John Dewey's *'The Quest for Certainty'*, although then I
had no clue who these writers were and how their position was in the field. I
think I should, at that time, have been made to study Nagel's very interesting
and illuminating chapters on the methodological problems of the social sciences and humanities that are, he clearly explains much less different from
those in physics then generally believed. The reviewer in *The Times Literary
Supplement* thought these chapters were *'the most interesting in the book'* as
Nagel *'is concerned to establish that the social sciences are capable of producing useful general laws and explanations though their methods are necessarily not completely identical with those of the physical sciences...For the
defense of the social sciences he considers among other, the objectives of
non-repeatability and subjectivity in the selection of materials.'* Unfortunately,
as said these chapters were exempted from my examination and only very
recently when preparing for this writing I returned to Nagel and read them
45 years too late. Only very recently I realized that professor J.J.A. Mooij, a
scholar of mathematics, physics, ethics, literature and analytical philosophy,
who was the examiner, like Nagel, probably must have had affinity with
American pragmatism, especially Peirce and must have also known Toulmin's
The Uses of Argument from 1958. Apparently, I was well primed by this pre-
parative reading, as I received Polanyi's *Knowing and Being*, as a gift from
close friends in February 1976 on the occasion of my BSc graduation. In
Polanyi's book the piece on *The Republic of Science* and comment on
C.P. Snow's *The Two Cultures* are still quite amazing (Polanyi & Grene,
1969). I then bought Polanyi's *Personal Knowledge* in July 1976. Despite my
disagreement with Polanyi's ideas about the interaction between science and
society, for me his work really was an eye opener presenting intuitive and
pragmatic support for the new post-empiricist philosophy (Polanyi, 1962). On
my shelves I still have also one of the books of C.A. van Peursen, *Wetenschappen
en Werkelijkheid* published in Dutch in 1969, which I read and marked up in
the fall of 1975 preparing for the course. Van Peursen, who was a leading
philosopher in the Netherlands in his time, already concluded that the best
philosophy of science was a mix of Popper's and Dewey's philosophy, also

(continued)

referring to the later work of Wittgenstein, Quine, Polanyi, Winch, Gadamer and Habermas. At the end of this book he critiques the idea of value-free inquiry and with Dewey and the pragmatists firmly states that scientists, here used as including scholars in SSH, don't need to complement their work with 'diepzinnige' theories about 'reality'. 'Diepzinnig' may be translated with 'profound', but also with 'abstruse' or even 'esoteric', and it is the latter word that Dewey used to criticize philosophy which in his opinion had lost touch with science and the real world. Science and knowledge, he states was not the goal, but that science and research are integral to the life we live and want to live and are an important means to the end of our responsibility to create instruments for the right policies and their actions. In August 1976 I bought *Technik und Wissenshaft als Ideology* by Jürgen Habermas which made a huge and lasting impression (Habermas, 1968). Habermas argues for an ethically and politically proper interaction between science and social life and offers a model for it that is explicitly based on Dewey's pragmatism. My recent revisiting of this early work of Habermas made me realize that the discussions in those days about Science and Society took place in a very different public context than the current discussions about Open Science. Yet, the message to opening up science and engage and communicate with the publics is the same. Finishing this book in the early summer of 2020, I must hopefully add that the COVID-19 pandemic has made a lot of people in science, society and government aware of the power of the practices of Open Science. As the corona crisis was not only a global public health catastrophe but also caused a deep global social and economic crisis, the idea that we can do science differently may even linger a bit longer than it did after both world wars.

As has been noted by many, the very first lines of Kuhn's book immediately disclosed the exact same problem, I here in this book still feel must be addressed, although in 2020 for slightly different reasons: *'History, if viewed as a repository for more than anecdote or chronology, could produce a decisive transformation in the image of science by which we are now possessed'*. At few lines down he states: *'This essay attempts to show that we have been misled by them in fundamental ways. Its aim is a sketch of the quite different concept of science that has emerged from the historical record of research activity itself'....however this new concept of science will not be forthcoming if historical data continue to be sought and scrutinized mainly to answer questions posed by the unhistorical stereotype drawn from science texts. ...a concept of science with profound implications about its nature and development (p1)* (Kuhn, 1962).

It was immediately very clear that Kuhn dramatically changed the discourse of the philosophy of science and its research agenda by taking the 'historical turn'. Ian Hacking has in his typical and eloquent but straight forward manner described the

conceptual differences between Kuhn and the major concepts commonly held in the
standard image of science (p6–16). (Hacking, 1983) These differences do concern
issues of how science is being done in the real, but also affect the philosophical
assumptions and prescriptions of the *'unhistorical stereotype'*. Differences do
regard the classical image of individual inquiry compared to communities of inquiry
bound by research traditions and paradigms and the idea of distinct phases of normal versus revolutionary science. The community aspect was not disputed, but a lot
of subsequent modern historical work showed that the very distinct scientific revolutions in time, as described by Kuhn in physics and chemistry, are not common and
that most of the time in science different schools and paradigms do operate simultaneously until one of them is favoured. Kuhn's work did not provide support for the
use a general method which unifies science, an important aspect of the standard
image of science for the positivists but also for Popper until then. But there was
more. A paradigm in Kuhn's view is a composite of classical internal formal scientific rules, techniques and experimental methods and values, but also conveys values
of external social, cultural, ethical and practical origin. These are involved in daily
question on which grounds new results and claims are judged by peers and when
major claims and theories are questioned, and their novel competitors have to be
considered. Paradigms give guidance in deciding what to belief. Here we advance
to the second level of criticism of the Legend. Kuhn based on his historical work
deviates from the positivistic norm of what scientific statements are, the analytic-
synthetic dualism and the criterium of objectivism, a major pillar of the Legend and
the Standard Model, as we have seen above. He was, a bit to his own surprise,
caught in serious long-lasting discussions about relativism, subjectivity and objectivity. These discussions about the internal logic and consistency of the major theories and assumptions of the standard image of science were in 1962 already for quite
some years ongoing between highly esteemed members of the discipline of analytic
philosophy, as we shall discuss below. Hacking wrote a very concise and compre-
hensible explanation of the immense importance Kuhn's book has had and still has
(Hacking, 1983) Kuhn did not only question the Standard Model and Legend
regarding the ideas about the scientific method versus its mismatch with the daily
practice, but he also questioned the logical-positivistic ideas of rationality. He did
not engage in their highly esoteric and technical discussion but showed based on his
historical work that scientists simply did not comply with some of the major prescriptions, and that anyhow even if they would have tried, they fail because these
could not be followed in the practice of inquiry. He receded to some degree in this
in response to his critics saying that he believed that internal empirical scientific
data and findings ultimately were the most important criteria for believing or rejecting a claim, statement or theory. It is of interest to note that after Kuhn's book
appeared 'fresh interactions between philosophers and historians of science' came
about. There may then have been several reasons for the separation of these now
closely related disciplines, but Toulmin very critically points to *'George Sarton
from Harvard (who) ruled over academic History of Science in the United States'*
and had declared collaboration taboo (p6). (Toulmin, 2001) Toulmin makes it clear
that the study of the history of science stood in a lower rank than philosophy and

that the history of science field had its own ideas about what good history scholar-ship was. With Kuhn, he concluded that historians held their distance from inquiry that involved study of external, social and cultural, economic and political factors. Bernal's seminal work *"The social function of Science* published in 1939 also for that reason was neglected for a long time (Toulmin, 1977).

The Empirical Turn in the Sociology of Science

The Other Mertonian Thesis. It is not only fair to say, but highly relevant for the logic of my book, that I until now presented the dominant and legendary interpreta-tion of Merton's sociology. This was the image of an autonomous social system which was governed in an ideal fashion by scientists who were not troubled by the moral and social defects of all other human beings in modern societies. But there was another side of Merton's sociology which is in agreement with the sociology of science that became mainstream in the 1970s but is of totally different kind as the Mertonian legend. Steven Shapin, and later Harriet Zuckerman, the latter who at Columbia was a collaborator of Merton and much later in life his married wife, demonstrated that Merton clearly recognised external influences on science and not only of the religious, but also of the utilitarian and military kind (Shapin, 1988; Zuckerman, 1989). Merton has become widely known, and criticized, for his thesis, following Max Weber's well know theory, that Puritanism, Calvinism and Pietism are important external factors that may explain why the rise of modern science occurred in Western Europe (Cohen, 1994). Shapin quotes many lines and phrases from Merton's early book on the history of science that was published in 1938 (Merton, 1938), to show that Merton has not been properly read in this matter: *'Merton then proceeded to point to "further orders of factors," some cultural, some social, that might be thought relevant to explaining the historical materials with which he was concerned. These included interesting speculations about population density, the rates and modes of social interaction characteristic of different societ-ies, and other features of the cultural context not included in religious construct. Merton carefully noted that Puritanism only "constitute[d] one important element in the enhanced cultivation of science." In other settings "a host of other factors - economic, political, and above all the self-fertilizing movement of science itself'-worked "to swell the rising scientific current." Since science burgeoned in Catholic sixteenth-century Italy, Merton freely acknowledged that "these associated factors" might come to "outweigh the religious component'.* p595–596. (See Shapin for ref-erences to these citations of Merton.) Merton describes the mutual interdependency of science with other social institutions and their vested interests which has directly or indirectly influenced the direction of science and research through problem choice. This obviously is a problem in view of disinterestedness and objectivity of the Legend, which Shapin addressed upfront: *'at the very core of his enterprise, historians nervous about the black beast of "externalism" should be reassured. Neither in his 1938 text nor in subsequent writings was Merton ever concerned to* *adduce social factors to explain the form or content of scientific knowledge or scientific method'*. p594 (Shapin, 1988). Merton discusses the external socioeconomic effects on the dynamics of problem choice and subsequently that of scientific (sub) disciplines. Issues of the different personal motivation's scientist may have and which they often openly state which may relate to the potential practical and technological application of their research but also looking to the social status of research for their upwards social mobility. These studies about social interdependencies seem to have been collectively and selectively overlooked by historians and sociologists, verging according to Zuckerman on counterfactual history. For Merton, as Zuckerman points out, during his whole career the Puritanism Thesis was minor, compared to *'military, economy, geography and society'* as is reflected in the number of chapters devoted to them in Merton's book of 1938, reprinted in 1970) and subsequent writings. She refers to I.B.Cohen's review of the book (after it was reprinted), who thought that this minimal interest in influence of socio-economic and military factors on science was in the 1930s not new because it was already a major theme in Marxist sociology of science, whereas the proposition of a connection between religion and science was novel. I argued above discussing the work of Bernal, in agreement with Shapin, that indeed these ideas were dominant in Marxist sociology and theory of science, but not acceptable outside these circles and surely not mainstream in the late 1930s. With McCarthyism in the late 1950s and after Sputnik, during the years of the Cold War these chapters on external factors were, to put it mildly, tainted with Marxism and Socialism and not 'in sync' with the ideologies and images of science of the Legend.

In the 1970s and 1980s a new sociology and history of science was developed, called Sociology of Scientific Knowledge (SSK), from the perspective that in a *'sociological approach to knowledge-making, people produce knowledge against the background of culture's inherited knowledge, their collectively situated purposes, and the information they receive from natural reality.'* (Shapin, 1982). This research in sociology and history thus goes further than the classical dominant forms of history of science and further that Kuhn by bringing in external social values in the equation. It not only, as discussed above, shows how the practice of science really is, but is also shows how theory choice is done and how beliefs and scientific statements become accepted, and in that respect provide empirical sociological evidence against the Legend. The quote above is from an early seminal paper by Steven Shapin, a historian who became in his own words a sociologist and was one of the pioneers leading the way in this new interdisciplinary field between history, sociology and to some extend philosophy of science. Shapin very explicitly contrasts the two main approaches to the study the sociology of scientific knowledge. I will stay away from too much technical language but summarize the main points most relevant for the context of our present discussion of the demise of the Legend. Shapin builds a strong case, with a well-developed critique of the mainstream history of science, complemented by an overwhelming series of examples of more recent historical research with an empirical sociologically inclination. The latter research by among others Collins, Pickering, Geison, Wynne, Harvey, MacKenzie and Barnes, and Latour and Woolgar produced evidence obtained from cases widely distributed in time, place, and discipline for influences of 'non-cognitive' external cultural and religious values, political principles, beliefs and ideas on the process and the ultimate outcome of scientific inquiry. In effect, supporting the theoretical hypothesis as formulated in the quote on top of this paragraph.

We have seen above that the dominant history of science before 1960 or so, was confirmatory to the myth of the Legend and positivism under heavy direction of George Sarton. In a striking analogy, also in the history of sociology such a thing has been dominant for a large while. I will cite Shapin on the characterization of this sociology which he calls *'the coercive model'*. I will start with his conclusions: *'.... more significant problem arises from a largely informal model of sociology of knowledge which seems to be prevalent among a number of philosophers and historians of science.....Its main characteristics can be briefly described: (i) it maintains that sociological explanation consists in claims pf the sort "all (or most) individuals in a specified social situation will believe in a specified intellectual position"; (ii) it treats the social as if one could derive it by aggregating individuals; (iii) it regards the connection between social situation an belief to be one of 'determination' although little is explicitly said about the nature of determinism; (iv) it equates the social and 'irrational'; (vi) it sets sociological explanation against the contention that scientific knowledge is empirically grounded in sensory input from naturally reality.'* This has informed the classical sociology of science with respect to the role of individuals in the community *'generally regarded as troublesome'* and *'the connection of the social and the cognitive would generally be sought through the use of individual orientation particularly through motivation...factors internal to the scientific community would be viewed as non-social. Finally one would say as little as possible about the fact that scientists conduct experiments, look through microscopes, go on field expeditions, and the like , for wherever 'reality enters in, the sociological explanation is obliged to stop ...the coercive model has* **two splendid advantages**. *First...no successful instance of its practice will ever be encountered. Second, it portrays the role of the social and of sociological explanation in an unpalatable normative light: as if it were said that "no rational person would ever allow himself to be socially determined! Nevertheless, there is one major problem...; namely that it is not an accurate picture of sociological practice'. P195 In a sociological approach to knowledge-making, people produce knowledge against the background of their culture's inherited knowledge, their collectively situated purposes, and the information they receive form natural reality. Perhaps the most puzzling charge sometimes laid against relativist sociology of knowledge is that it neglects the role played by sensory input. On the contrary, the empirical literature employing this perspective shows scientists making knowledge with their eyes wide open to the world' p196.*

Shapin explicitly elaborates on inquiry and its purposes and goal-directedness not set by *'contemplative'* individuals, but by a community where *by doing things with knowledge that its meaning is produced'. The purpose for which knowledge is produced and according to which it is evaluated may vary widely: they may include legitimation or criticism of tendencies in the wider society, or they may encompass goals generated exclusively within the technical culture of science.'*

Shapin argues that the ideal type of the modern scientist should take these sorts of considerations, of this broader spectrum of social and cognitive scientific interests, into account. In this view of science, which is not compatible with 'rationality' of the normative Legend, according to Shapin, *'the role of the social is to pre-structure choice and not to preclude choice'* p198.

There clearly is in 1982 still a huge tension here with the Legend and its positivism: *'While it may be banal to say that statements of scientific fact may be theory-laden. It is not, apparently banal to* <u>*demonstrate*</u> *this empirically and to pin down the specific networks of expectations and goals affecting the production and evaluation of statements of facts...Historians act as if, after all, observed facts count as 'hard case'; making a fact into a historical product (an artefact) is an exercise which historians of science approach with great caution (even though scientist do it routinely)'* (p159). The latter remark is of interest and sounds familiar in the present context because it refers to the way how active scientists 'pragmatically' deal with these philosophical ideas. Shapin states that the classical historians of science assumed that with the professionalization of science the scientific community obtained autonomy towards social factors and their influences. Here social factors are regarded as limited to obviously external social and political values. *"To many writers an 'influence from Malthus (or from Paley)* [on Darwin] *has not been something to describe and explain, but something to be* <u>*explained*</u> *away, since from the present perspectives it would be regarded as an illegitimate inclusion in properly objective scientific thought.'* It is because of this influence, according to Gillispie, *'that it is inconceivable that* the *Origin of Species could have been written by any Frenchman or German or by an Englishman of any other generation.'p179.*

Shapin draws attention to professional vested interests that are internal to science and research, but not strictly cognitive. Active scientists know these very well as they determine the ongoing discussions, at the moving front of research, with reviewers 'from other schools' at journals, grant review committees, scientific committees selecting conference contributions (selection of main speakers and of oral abstract presentations), academic promotions committees and decisions who writes or contributes to textbooks. All of these judgements determine what 'we' hold to be 'good' research or 'the best' research at some point in time, which over my 40-year career developed and changed rather quickly (Miedema, 2012). An outstanding analysis of the diversity of private, professional, cultural, social and economic factors that influence the practice of inquiry and knowledge making is Gerald Geison's study *'The Private Science of Louis Pasteur'*. (Geison, 1995) This book was by many especially French scientist considered to be debunking Pasteur. It was published, at the same time as more hagiographic biographies at the centenary of his death in 1995, but by experts highly praised because it provides deep and detailed insights how knowledge, in basic but also in applied biomedical research with enormous societal and economic impact, was and is produced. In a critical, humiliating review of Geison's book, Max Perutz (1914–2002), who was a famous biophysicist, defended Pasteur, against Geison's demonstrations and judgements of Pasteur's obvious foul play (Perutz, 1995). The bottom-line of the defence was that in the end Pasteur had been proven right, only the facts count in Perutz' opinion. The real issue at stake, that clearly surfaced in the exchange that followed in the NYRB, was that an outsider, not a man or woman from the lab, apparently not with 'pious reverence' and excessive respect, was messing with men of science and its methods (Miedema, 2012). Other writers of recent history of science, such as Crewdson have been overly critical, for instance regarding the role of Robert Gallo, in a study of the discovery of HIV, the aids virus in 1983 (Crewdson, 2002). Crewdson on the other hand has undue sympathy for the 'underdog' in this dispute that involved massive professional reputation including a 2008 Nobel prize, national politics and economic interests (Miedema, 2002).

Shapin has since 1982, written a number of classical highly influential journal articles and books about the practice of science and the production of knowledge in the seventeenth century and in our times by doing in depth research using historical and sociological methods in which all of the above topics, theories and problems are addressed (Shapin, 1994, 1996, 2008; Shapin et al., 1985) In the last pages Shapin provides a balanced discussion of how to view the influence of external and non-cognitive factors on knowledge production. Some researchers simply regard it as wrong based on the ideal of objectivity and value free science and studies in sociology or history that reveal these influences are considered damaging and 'aspersions'. Some regard these influences as realistic, it happens and is difficult to avoid, but they are per definition corrupt because science is, and its institutions are in that way being hijacked by all kinds of powerful politically and socially organized groups and their interests. Shapin regards these views as 'a misunderstanding' as external values and concepts have had and may have beneficial effects on the growth of knowledge. Opening up science to less powerful publics has these risks and as discussed in depth in Chap. 5, it will require continuous debate to resist the capture of science by the economic powers in society.

The Myths of Science: Frontstage and Backstage

Humans and scientist alike need certainty, a logical method, an algorithm, with timeless and thus objective foundations. But the Quest for Certainty has failed. We have in reaction to this in the 1990s seen academic debates and worries about loss of certainty and foundation of scientific truth. This mainly was a reaction against certain forms of excessive post-modernism, relativism and subjectivity. Several authors have discussed these worries to demonstrate that science is unique as a knowledge producing system, that produces robust, reliable and significant scientific knowledge even if we acknowledge that there is no metaphysical, given formal method or rules and foundations to guide us at truth. I will return to that discussion in Chap. 4 when discussing the default of pragmatism for the philosophy of science after the era of the Legend.

For now I want to discuss the reasons for the anxiety and worries academics experience whenever it is publicly discussed that the legendary image of science does not match with the practice behind the doors of the sociology, psychology, philosophy and history departments, but don't make a mistake, behind the doors of laboratories of the natural, biomedical and geosciences as well. This anxiety almost every time pops up also at less public debates about the Legend and how to arrive at a more inclusive way of thinking about science and the design and organisation of our academic institutions. I use the vocabulary frontstage – backstage from a framework developed by Goffman (1959). Thinking about the Legend, our popular image of science, the myth of which has been shattered by its novel criticisers but also by its erstwhile major proponents, Goffman's dramaturgical model for social interactions can be of use. Not only humans in their interactions knowingly assume different behaviour and roles regarding the relationship, interaction and social context they operate in, but likewise public organizations and institutions show different behaviours in different situations, meant for different publics. In many instances in public theatres, formal meetings or media appearances presentations by representatives from financial institutes, banks, government or institutes affiliated with government, the church, the hospital administration and private companies follow the frontstage narrative or storyline. This, of course, presents the perspective of a reassuring, sophisticated, empathic, politically and socially correct reflective organisation. Of course, for different organizations, different items may be considered for an idyllic frontstage story and attitude. It is precisely this function that the Legend has had, and to a still lesser degree still has for science and academia. Most of the writers I cited thus far and will be cited further on, in the introduction and epilogue, but often throughout their analyses in many different wordings relate to the worry they or the scientific community may have when they debunk the myth of the Legend. The myth of the Legend, as demonstrated above, has been debunked by a few in the 60s, but openly many times since the 1970s by prominent thinkers which has reached a relatively wide audience, outside and inside science. Relatively, since in most cases even during the so-called Science Wars of the 90s when a larger audience got interested in a short while, it is a fairly limited readership. As pointed out above, active natural science researchers or even humanities scholars, in normal times take the Legend for granted, they intuitively know how to produce knowledge and now the mores of their field, but get nervous when the spell is broken, the myth of the Legend destroyed. All of a sudden one has to realize what the real backstage situation and the correct corresponding narrative is for that. That is very, very hard, since we are coming from the Era of the Legend, where scientific inquiry as we have seen is held to be unique, timeless, to provide for knowledge with absolute truths and because of its methods, rules and bedrock logical foundations has proven to be successful and to be successfully applied in the modern world. It may have been a problem for the philosophers who gradually saw the Quest for Certainty and their dreams and wonderful philosophies of timeless foundations, unified science, formal analytic methods, realism and positivity come to an end. For scientist and those working in science and academia the problem is less esoteric and practical but felt to be tricky. The fact that we can no longer use the Legend as a frontstage ideal narrative of science, that has carried much weight since the 1940s, is indeed difficult. It has been rather effectively used to claim authority for science in public debates, about safety of vaccines, the cause of climate change and what should be done about that. It has been used to discuss many public health and prevention and political issues relating to inequality, fair economics, the regulatory role of government in neoliberal times, but also on an annual basis by some about the absolute prominence of basic natural science.

So, one wonders if we admit that science in the real is done as we do it -producing the claims and insights we believe by a uniquely robust and open, continuous purging, process of testing, of experiments, repeating of experiments, a lot of criticism and debate in a cycle of improving and rejecting- will that convince the public as well as we did convince them with the story of the Legend? Most of the writers, including myself, say yes, that shall do. Be honest, show how knowledge is and has been produced, how robust the process is also when we know that social interests of cultural and personal source are at play. Be frank about the fact that every claim, theory, method, action based on this process is fallible and may eventually be improved, corrected and rejected because it is replaced by a better alternative.

I will here not discuss the Science Wars of the 1990s. *'The One Culture'* by Labinger and Collins (2001) presents a highly readable series of short papers of heavily involved authors with different perspectives on that. The Science Wars was a reaction of the natural science defenders of the Legend to claims in academia that-because postmodernist relativism had shown that there is no scientific method as held by empiricism and positivism- scientific theories and accepted beliefs are in essence not different from the beliefs derived outside science from superstition and all kinds of popular, religious and personal opinion. This image of science, which derives to be honest in some respects from Rorty's bold interpretation of Willem James' pragmatism, was at the far end of the spectrum opposite of the theory of inquiry of Peirce and Dewey, later extended especially by Putnam (see Chap. 4). The defenders or 'bulldogs' of science went all out with an appeal on the Legend which was not constructive. Fortunately, many philosophers have offered realistic and pragmatic views of science and its practice, without taking refuge to metaphysical and foundational myths of the Legend. I refer to Ian Hacking again, and especially his *The Social Construction of What?* where he in Chaps. 1, 2, 3 and 7 makes a very clear case for the realistic and naturalistic middle ground (Hacking, 1999) and to a very insightful and opiniated review by Shapin (1982). These studies show that there are clearly social and cultural factors at play, but that there are constraints to our claims and ideas in the confrontation with and observation of natural and social reality and these together in a continuous critical debate guide the process of how beliefs get accepted and hypothetical claims become facts. (p33) Our realistic understanding of the practice of scientific research, where collective reason, experiment and action ground our beliefs which is constrained by conditions in the real world being the natural or the social world.

The good news in my mind is that we can pragmatically make a very good point for the reliability of science as follows. Since modern times we have this new robust collective way of doing science by hypothesizing and experimentation, its ever-improving methods, techniques, technologies and the ever-growing collective experience with judgements of claims and experimental results using ever improving sophisticated methods and methods of reasoning. This has resulted indeed in impressive success, changing our lives by changing the unfriendly environment, improving our health and life expectancy, allow quick, convenient and mass transportation, modern communication, increasing personal and global wealth, dealing with issues of energy, and so on. This all has been achieved despite the fact that even in the natural sciences we never had a unified formal objective value-free scientific method, an no timeless foundation for our knowledge to build on. Social and political values have always at several levels been involved in our evaluations and criticism of what to study and what to belief in scientific inquiry. This inclusive deliberation has steered science in society also in modern history to the good but sometimes to the bad. Our common-sense collective methods of inquiry have brought us time and again wonderful results that changed our life's in the past 200 years.

'OMG.......There Is No Foundation!'

The epistemic core in the philosophy of science and the Legend is empty, was the conclusion of Nowotny et al in their *Re-Thinking Science* that I will discuss in detail in Chap. 5. But I use it here for its analogy with the evolution of the thinking that many of us have had regarding religious beliefs. The story about the Legend of science feels, I image, to many who were raised on the Legend since elementary school, high school and university as loss of certainty and loss of a familiar story that provides for calm and rest of mind. For me it compares to my growing up in a Calvinist family in the North of the Netherlands during the 1960s, where despite the non-academic background of our parents for them and us, reading and studying was part of life. Gradually, I came to realize at the age of 6, I think, that Santa Claus did not exist, but that was alright with me. Much more complicated was in the years between my 14th and 20th year how to think about the origins, foundations and the revelations of our Christian beliefs, ethics and ideal practices. Specially my father was convinced and believed the factual truth of the New Testament, from cover to cover, and this and the ethics and prescribed practices were regularly discussed at home. As a bachelor student I started reading modern theology amongst others Rudolph Bultmann, which made a lasting impression on me. In particular, his demythologization of the biblical texts and his rejection of the supranatural as world views belonging to another cultural context in the past, not appropriate for our modern time were strong images for me. He posed the idea that the biblical stories are not facts but language and texts describing acts of God. There is a core in the text, a message that in every time and culture can have its own narrative form. I had concluded that I did not believe in any of the supernatural, which until now has not caused me more than average anxiety. I was, however, for ever a Calvinist engaged by the ethics and social-democratic politics that came with my upbringing and later reading the modern ethical and political interpretations of the Biblical texts by

(continued)

Bonhoeffer, Sölle, Bloch, Moltmann, and Pannenberg. These writers influenced Kuitert a Dutch theologian who's public intellectual and emotional struggle I with many others followed since 1971 until his dead in 2017. In a series of books, he goes through a sequence of phases in which he gradually peeled of the layers of classical Calvinist theology and its dogma's. Eventually and it seemed inevitably, he had to admit in the 1980s that there was no foundation, all our speaking and theology about the divine and the supranatural was the product of humans. He was also clear to point out that these revelations thus were not Divine revelations and not God given. Here again the same question as for science comes up, do we have a good enough narrative about Christianity and religion in general if we demythologize its foundation and reduce it to ethics and action in contribution to human flourishing and the good life. Harry Kuitert argued that these 'inspired' ethics and this social-political awareness based on diverse cultural and personal values may shape socialist, conservative or liberal worldviews and policies alike.

2.2   Part 2. The Crisis in Analytical Philosophy

The spirit of Cartesianism is evidenced not only by rationalists but by all those who subscribe to strong transcendental arguments that presumably show us what is required for scientific knowledge, as well as those empiricists who have sought for a touchstone of what to count as genuine empirical knowledge.....the first attack was made by Peirce. Nevertheless it has taken more than hundred years for us to become fully aware of how the Cartesian view distorted the way in which science is actually practiced.'p71 (Bernstein, 1983).

The crisis in analytical philosophy started around 1960 in the philosophical discipline that created the problem in the first place. Crisis became apparent in open debates when philosophers officially declared the dead of positivism and empiricism. Philosophers had admitted much earlier that there were already cracks in the idea of a foundation and other aspects of the Legend. C. S. Peirce was on one of the first *'to attack the Cartesian framework especially in regard to characterizing scientific knowledge' (p71)* (Bernstein, 1983). His work, in the last decades of the nineteenth century that was followed up by the American Pragmatists James and Dewey until 1940, did not belong to main-stream analytical (logical-empiricist) philosophy and did not get much attention there, apart from Frank Ramsey who's engagement with pragmatism was cut-short and almost forgotten by his untimely dead in 1930 and Nagel, which will be discussed below. Eventually the debate developed with the work of W. V. O. Quine in the 1950s; Popper and Michael Polanyi 1958, 1959; Kuhn in 1962; followed by Toulmin, Feyerabend, Apel and Habermas, Hesse, Hacking, Putnam, and Rorty in the 1970s early 1980s.

This critique on logical positivism and empiricism in the 1970s reached a much larger audience also outside the departments of the philosophy of science. Gradually, it was picked up some active natural scientists or SSH scholars who had an interest in philosophy and sociology of science. However, it appeared -and even in 2021 appears- to be hard for several reasons to go beyond the truly mythical Legend, letting go of the ideas of a timeless foundation and the dreamed formal methods of a science, even when it was realized that it was a method of 'a science that never was'.

Regarding what was at stake, I will again quote Bernstein who has discussed in great detail and transparency these debates and in strong statements the image of science that emerges in the post-empiricist philosophy of science in contrast to image of the logical-empiricists for which Ziman and Kitcher coined the name 'Legend': '*We can interpret this movement of thought as contributing to the demise of Cartesianism that has dominated and infected so much modern thought. The Cartesian dream of hope was that with sufficient ingenuity we could discover, and state clearly and distinctly, what is the quintessence of the scientific method and that we could specify once and for all what is the meta-framework or are the permanent criteria for evaluating, justifying, or criticizing scientific hypotheses and theories. The spirit of Cartesianism is evidenced not only by rationalists but by all those who subscribe to strong transcendental arguments that presumably show us what is required for scientific knowledge, as well as those empiricists who have sought for a touchstone of what to count as genuine empirical knowledge.....the first attack was made by Peirce. Nevertheless it has taken more than hundred years for us to become fully aware of how the Cartesian view distorted the way in which science is actually practiced.'p71* (Bernstein, 1983).

A Detailed History of the Philosophical Demise of the Legend

As I pointed out at the introduction of this chapter, it is the analysis of the origins and effects of exactly this distortion that is the topic of this book. I will in these remainder of this Chapter discuss the philosophical arguments that convincingly show why the analytical and positivist philosophy failed. I will not go in great detail about the technical discussions. I chose to offer a diverse chronological selection of thoughts and conclusions of the most prominent scholars. I provide the most illuminating citations taken from their work. Readers may wonder why I sometimes cite longer paragraphs. It is because in my opinion they are essential and because I want to give the reader the opportunity to directly read this primary 'material' with no need to have to rely on and trust my paraphrase's and interpretations.

Below I discuss in historical order the work of the major scholars since 1945 the problems of positivism, the analytical philosophy and empiricism, which demonstrates the collective developments in the field and in some cases the personal development and struggle to break free from foundationalism. For the readers who do not know the authors which I am going to name in the remainder of this chapter, without exception they all were, or are when still alive, the absolute top scholars in their field. It makes you wonder that only the true elite, the leading scholars in exactly the field of interest were in a position to challenge the main theoretical ideas and concepts of logical positivism and empiricism, largely the legacy of the Vienna Circle that had been build up over the past 50 years. Most of them had actually trained with that previous generation of top philosophers who had all contributed and shaped exactly these philosophies. They were mostly students or second-generation students of Wittgenstein in the UK, and Carnap, Reichenbach, Hempel, Quine in the USA. It apparently is quite difficult, and it requires a reputation and a position of intellectual power to change the thinking in a field, which is a case in point of Kuhnian paradigmatic revolution and of the power struggle in a 'field' as described by Bourdieu. (Bourdieu, 1975) Ludwick Fleck, who anticipated Thomas Kuhn's major work by at least 30 years, writing in 1935 about criticism in science said that writers who trained as sociologist or in classics, 'no matter how productive their ideas, commit a characteristic error (Fleck, 1979). They exhibit an *'excessive respect, bordering on **pious reverence** for scientific facts'*, cited by Ian Hacking, but Hacking adds: *'The era of excessive respect has passed'* (p60) (Hacking, 1999). I believe that this excessive respect was not so much for 'scientific facts', but for the mythical power of the scientific method of positivism that claimed the status of these facts and the status it provided to the scientists.

I start with C. S. Peirce who wrote long before any of them and was part of his own "Metaphysical Club' some 30 years before the Vienna Circle had started (Menand, 2001; Misak, 2013a). Peirce, as said, was later recognised as the *'first to attack the Cartesian framework'* and influenced many if not all of major modern philosophers, before 1940 and after some lag time again directly or indirectly since the 1970s (Bernstein, 2010).

C.S. Peirce who did his most influential writing at the end of the nineteenth century, was one of the first to attack the Cartesian framework (Bernstein p71). The framework of the idea of a transcendent foundation and the empiricist formal method. He was trained as a chemist in the natural sciences and is now considered to be exceptional regarding his many original ground-breaking contributions to natural science and in particular in the philosophy of science which are studied with renewed interest until this day. Many influential philosophers have payed tribute to Peirce.

Ernest Nagel in 1939:

> *'Peirce's distinctive contributions to logic as the general theory of signs, centre around his pragmatism, his critical commonsensism, and his fallibilism. By far the best known is his pragmatic maxim, proposed as a method for clarifying ideas, eliminating specious problems, and unmasking mystification and obscurantism hiding under the cloak of apparent profundity. In one form or another his proposal was adopted by a number of distinguished thinkers, for example, in this country by William James and John Dewey, so that to-day it is almost a common- place. Peirce's own formulation of the pragmatic maxim leaves much to be desired in the way of explicitness and clarity; and more recent formulations, such as those by Professor Carnap and others, have the same general intent but superior precision. I nevertheless venture two general remarks on the Peircean version of pragmatism which, though obvious, merit attention. The pragmatic maxim was intended as a guiding principle of analysis. It was offered to philosophers in order to bring to an end disputes which no observation of facts could settle because they involved terms with no definite meaning. It was directed at the Cartesian doctrine of clear and distinct ideas, which found the terminus of analysis in vague abstractions claimed to be grasped intuitively, as well as at the common tendency to convert types of behaviours into unknowable agencies controlling the flux*

of events. Above all, it pointed to the fact that the "meanings" of terms and statements relevant in inquiry consist in their being used in determinate and overt ways. **Pragmatism, to employ Peircean language, was thus a proposal to understand general terms in terms of their concrete application, rather than vice-versa.** *'(bold case by FM)*

At the risk of treading upon ground on which angels fear to step, I should also like to mention the elementary point that in terms of Peirce's emphasis neither terms nor statements can be regarded as designating, independently of the habits involved in their use. Consequently, "the meaning" of expressions is not to be sought in self-subsisting "facts", "essences", or other "designata", but must be construed in terms of the procedures associated with them in specific contexts.

'Peirce claimed no infallibility for the beliefs of every-day experience, and indeed one of the cardinal tenets of his thought was a universal fallibilism. Peirce's fallibilism is a consequence of his regarding the method of science as the most successful yet devised for achieving stable beliefs and reliable conclusions; it has nothing to do with the malicious scepticism which rejects science on the ground that its conclusions are after all not established as being beyond the possibility of error, only to invoke a special set of imperatives as indubitable objects of human endeavour. Peirce noted that the conclusion of no scientific inquiry is exempt from revision and correction, that scientists feel surer of their general logic of procedure than of any particular conclusions reached by it, and that the method of science is self-corrective, both as to its own specific features and the specific conclusions gained with it.' Read at the Fifth International Congress for the Unity of Science, Harvard University, September 3–9, 1939. (Nagel, 1940)

Habermas in his *Erkenntnis und Interesse* in 1968 translated in English in 1971 (ref) devoted two chapters to Peirce: *'What separates Peirce from both early and modern positivism is his understanding that the task of methodology is not to clarify the logical structures of our scientific theories but the logic of the procedure with whose aid we obtain scientific theories. We term information scientific if and only if an uncompelled and permanent consensus can be obtained with regard to its validity. This consensus does not have to be definitive but has to have definitive agreement as it goal....modern science distinguishes itself by a method of arriving at an uncompelled consensus about our views.'* p91 *'For Peirce there was one method of inquiry, based on deduction, induction and to a small degree inference to the best explanation* (designated abduction by Peirce). *Truth was roughly, whatever hypothesizing, induction and testing settled down on.'* (p118) Peirce named it the 'scientific method', the logic of or method of inquiry, but he did not mean to suggest that it is a logical formal system that allows us to get to the truth. Habermas: *'For Peirce this concept of truth is not derivable merely from the logical rules of the process of inquiry, but rather only from the objective life context in which process of inquiry specifiable functions: the settlements of opinions, the elimination of uncertainties, and the acquisition of unproblematic beliefs-in short the fixation of belief.'* p119 Peirce resolutely rejected the Cartesian foundations, transcendental necessity and conditions, the so-called 'spectator theory of knowledge' that assumes the fact-value dualism. *'For Peirce it is the method'*, says Habermas, *'that takes over the role of an unshakable foundation, the a priori judgements that per definition cannot be doubted because they are a 'given'.* p97 This thinking of Peirce was many years later followed up by great men like Sellars and Quine. Peirce assumed a constant state not of scepticism but fallibilism, with continuous doubt about our claims in which he anticipated much of Popper's falsificationism published in 1935.

John Dewey already concluded in the beginning of the twentieth century in many of his writings that philosophy appeared to be an internal debate for philosophers, esoteric and of little value to understanding and guiding the practice of scientific investigation and its relation to reality, society and human life. Dewey in *The Quest for Certainty* (1933) and elsewhere wrote extensively about what Bernstein called 'the 'Cartesian Anxiety', *the belief of Descartes that the philosopher's quest is to search for an Archimedean point on which we ground our knowledge'*. (1983, p16) I will cite the crisp and concise remarks of Hacking about Dewey's criticism of the philosophy of science of the empiricist and positivistic tradition. Hacking later confessed (Misak, 2007) that he himself found it hard to read Dewey, *'it goes on and on'* and that feeling is familiar to me. Hacking: *'Truth is whatever answers to our present needs, or at least those needs that lie at hand. Dewey gave us the idea that truth is warranted acceptability. The world and our representation of it seems to become at the hands of Dewey very much a social construct. Dewey despised all dualism- mind/matter, theory/practice, thought/action, fact/value. He made fun of the 'spectator theory of knowledge'. He said it resulted from the existence of a leisure class, who thought and wrote philosophy, as opposed to a class of entrepreneurs and workers who had not the time for just looking.*

Hacking, says about Dewey: *'My own view, that realism is more a matter of intervention in the world, than of representing it in words and thought, surely owes much to Dewey.'* (p62) (Hacking, 1983). Pragmatism, from Peirce, Dewey, James, Nagel, Quine to Habermas and Hacking, is beyond Cartesian empiricist philosophy and holds that it is this relation to practice, intervention, and actions based on our accepted beliefs that gives value to our beliefs, and not timeless transcendent formal principles that cannot be tested.

Karl Popper (1902–1994), was a most influential philosopher of science who in his later years also wrote extensively about the open society, freedom and democracy. He was in time and space close to the empiricist positivist philosophers of the Vienna Circle but did however not agree with most of their philosophy. In his "*Logic der Forschung: zur erkenntnisstheorie der modernen naturwissenschaft*" published in 1935, a translation of which appeared in 1959 under the title "*The Logic of Scientific Discovery*", he criticised the positivist and empiricist philosophy on their major ideas. It has been said that this critique, after the members of the Vienna Circle having tried to incorporate some of it, eventually in the 1950s caused the declaration of the death of logical positivism. Popper wrote in his autobiography, *Unended Quest* (chapter 17), that he rather thought the Vienna Circle came to end because they did not address the real problems, but got immersed in debates about minor problem, puzzles and in particular the meaning of words. Although this echoes the critique on philosophy of Peirce, James and Dewey, they are not mentioned by Popper in this discussion of philosophy of science. Toulmin, but not even Kuhn is mentioned which is remarkable, given the impact of Kuhn's work on the legacy of logical positivists, that was already tangible at the time of Popper's writing (Popper, 1976).

In the 1958 introduction to the English translation of *The Logic of Scientific Discovery*, Popper states that he is a pluralist and he commends the philosophers *'who do not pledge themselves in advance to any philosophical method, and who make use of epistemology, of the analyses of scientific problems, theories, and procedures, and, most important, of scientific discussions. …Its most important representatives… were Kant, Whewell, Mill, Peirce, Duhem, Poincaré, Meyerson, Russel and later in some of his phases Whitehead. Most of those …would agree that scientific knowledge is the result of the growth of common-sense knowledge. But all of them discovered that scientific knowledge can be more easily studied. It's very problems are enlargements of the problems of common-sense knowledge. For example it replaces the Humean problem of 'reasonable belief' by the problem of the reasons for accepting or rejecting scientific theories.'* p22 (Popper, 1959).

Hacking compared Popper's philosophy with that of Carnap's logical positivist philosophy, saying *'They disagreed about much, only because they agreed on basics. It would be nice to have a criterion to distinguish such good science from bad nonsense or ill-formed speculation.'* (p3) Hacking, who wrote that he has been most influenced in his early days in England by Popper, concludes that despite these differences the positivists and Popper contributed a lot of the timeless image of science The Legend, that ruled before Kuhn, before 1960: *'They thought that the natural sciences are terrific and that physics is the best. It exemplies rationality and from that they believed in the unity of science.'* p5 (Hacking, 1983).

As I have discussed above, the positivists started with observations from the bottom, building it up into a system of verified statements about the world. Popper did reject this idea on philosophical logical arguments. In his view it starts top down with hypotheses, that are based on previously obtained knowledge, discussions with peers or simply wild ideas. These conjectures and their contexts determine how we subsequently observe and how we interpret the observations about the world. In Popper's view the claims derived from these observations may after severe experimental testing and discussion between scientists become accepted, held to be 'true'. However, per definition they are not verified. On the contrary, theories and their statements are to be regarded falsifiable, open to refutation, at any time by further testing and criticism. Poppers 'method' of conjectures and refutations, and his falsificationism reminds of the 'scientific method' described by Peirce 50 years before. Like Peirce, Popper completely rejected the idea of the independent, 'given' foundation and the dichotomy between facts and values. Observation, ideas and theory were always entangled. In his thinking, like Peirce, Popper emphasized the power of the method of rigorous and endless testing and of criticism in the <u>community</u> of peers. *"Basic statements are accepted as the result of a decision or agreement, and to that extent they are a convention. The decisions are reached in accordance with a procedure governed by rules'….. 'Thus the real situation is quite different from the naive empiricist. Or the believer in inductive logic.'…. 'Theory dominates the experimental work from its initial planning up tot the finishing touches in the laboratory'. (p106)* In a most fascinating metaphor of the 'swamp' he resolutely deals with the issue of the foundation and the 'given'. It is the visualization of this powerful metaphor that I literally never got out of my mind after reading it in August 1975: *'The empirical basis of objective science has thus nothing 'absolute' about it. Science does not rest upon solid bedrock. The bold structure of its theories rises, as it were, above a swamp. It is like a building erected on piles. The piles are driven down from above into the swamp, but not down any natural or 'given' base; and if we stop driving the piles deeper, it is not because we have reached firm ground. We simply stop when we are satisfied that the piles are firm enough to carry the structure, at least for the time being'.* (p109).

In his 1972 Addendum he added: *'1. My term 'basis' has ironical overtones; it is a basis but is not firm. 2. I assume a realist and objectivist point of view: I try to replace <u>perception</u> as 'basis' by <u>critical testing'. Our observational experiences are never beyond testing; they are impregnated with theories. 4. 'Basic statements ...are like all language, impregnated with theories</u>'.(p109)* In a later paper, *The Rationality of Scientific Revolutions,* which takes in to account the community of inquiry and some of the sociological and psychological aspects of the research process, he describes the problems that may arise from this phase of debate and criticism due to the human factor (Popper, 1981) which were discussed in his *Conjectures and Refutations* at length (Popper, 1972).

Willard Van Orman Quine (1908–2000) is everywhere, when you read about the demise of the Legend and about his role, or not, in the resurrection of pragmatism (Misak, 2013b) Quine was familiar to the members of the Vienna Circle but worked whole his life in the USA. In most cases his contribution is very briefly told with short citations. He did not write much, but he made an immense mark through his famous dogma's on empiricism especially by forever rejecting, on analytic logical grounds, thus by using their own weaponry, the analytic–synthetic distinction. This was a blow to the very important yardstick of logical-empiricism and the philosophy of the Vienna Circle (Quine, From a logical point of view, 1956). He demonstrated, or in fact built the argument that the principles of inference that we use to link theory with experience [observations done via our senses] are as Putnam (Putnam, 1981) says [not analytical, nor given or timeless foundations but] *'are just as much subject to revision as any other aspect of our corporate body of knowledge.'* (p30). These rules are thus not 'given', or <u>a priori</u> assumptions but result from our collective thinking, experience and discussion and are such that as Misak phrases: *'everyone would assent to them'* (p200) (Misak, 2013a).

Michael Polanyi wrote in 1959 a short fascinating comment on C.P. Snow's *Two Cultures* that originally appeared in *Encounter*, a monthly Anglo-American journal of politics and culture that did fit Polanyi's political ideas discussed above. This piece is written in the characteristic polemic style of Polanyi who also here puts the issue in a larger political neo-conservative frame, criticizing the hard-boiled scientific ideals and naturalistic scientism of Bentham and Marx that in his view disrespects truth. *'Our task is not to suppress specialisation of knowledge but to achieve harmony and truth over the whole range of knowledge. This is where I see the trouble. Where a deep-seated disturbance was inherently originally in the liberating impact of modern science on medieval thought and has only later turned pathological'.. 'Science rebelled against authority. It rejected deduction from first causes in favour of empirical generalisations. Its ultimate ideal was a mechanistic theory of the universe, though in respect man it aimed only at naturalistic explanation of his moral and social responsibilities'. '..scientific rationalism has been the chief guide towards all the intellectual, moral, and social progress on which the nineteenth century prided itself- and to the great progress achieved since then as well. ...Yet it would be easy to show that the principles of scientific rationalism are strictly speaking nonsensical. No human mind can function without accepting authority, custom and tradition: it must rely on them for the mere use of a language. Empirical induction, strictly applied can yield no knowledge at all and the mechanistic explanation of the universe is a meaningless ideal....because the prediction of all atomic positions in the universe would not answer any question of interest to anybody'. 'Scientific obscurantism has pervaded our culture and now distorts even science by imposing on it false ideals of exactitude'. (p41)* (Polanyi & Grene, 1969).

Ernest Nagel has been an influential philosopher, not only through his famous textbook *The Structure of Science: Problems in the Logic of Scientific Explanation.* (Nagel, 1961) In that pre-Kuhnian seminal work he covered the whole of the philosophy of science of those days, but mostly limited to mainstream analytical philosophy, logical-positivism and empiricism and Popper's philosophy. There is a very short discussion of 'instrumentalism' which refers to American Pragmatism. He was sympathetic to pragmatism as I will discuss later and in his introductory chapter he makes a few remarkable statements which are a critique of the empiricist-positivist philosophy of the 'Legend' that he discusses in the next 300 pages of his book. *'The practice of the scientific method is the persistent critique of arguments in the light of tried canons for judging the reliability of the procedures by which evidential data are obtained and for assessing the probative force of the evidence on which conclusions are based'. ....'the difference between the cognitive claims of science and common sense which stems from the fact that the former are the products of scientific method, does not connote that the former are invariably true.'...... 'If the conclusions of science are the products of inquiries conducted in accordance with a definite policy for obtaining and assessing evidence, the rationale for confidence in those conclusions as warranted must be based on the merits of that policy. It must be admitted that the canons for assessing evidence which define the policy have, at best, been explicitly codified only in part, and operate in the main only as intellectual habits manifested by competent investigators in the conduct of their inquiries. But despite this fact the historical record of what has been achieved by this policy ....leaves little room for serious doubt concerning the superiority of the policy....'* (p18)

> *'For in point of fact, we do not know whether the unrestrictedly universal (positivist-empiricist premises) assumed in the explanation of the empirical sciences are indeed true... .'were this Aristotelian requirement adopted few if any of the explanations given by modern science could be accepted'... 'In practice it would lead to the introduction ...that explanations are being judged to have merit by the scientific* (p43)

Polanyi, who as we saw criticized positivism, concludes two different things from Nagel's account of science: *'Nagel implies that we must save our belief in the truth of scientific explanations by refraining from asking what they are based upon.*

Scientific truth is defined, as that which scientists affirm and believe to be true. Yet this lack of philosophical justification has not damaged the public authority of science, but rather increased it' (Polanyi, 1967)

Marxism? Critical Theory?
Before discussing Kuhn's work and immense impact from the 2020 perspective, I want here from the 1977 perspective refer to another writer who has until this day influenced my thinking about science, research and society. In September 1976, after a year of philosophy, I had returned to the lab bench to study for my Masters in immunology at the Academic Hospital of the University of Groningen. I continued reading about science and in the Spring 1977 I read Jerom Ravetz's book *'Scientific Knowledge and its Social Problems'* (Ravetz, 1971). Ravetz (1929-) is a mathematician who became a philosopher of science. After his graduation in the US, he came in the late 1950s to the UK at a time when his even moderate Marxist sympathies were problematic with McCarthyism in the US. In Europe Marxist sympathies in the 1960s and 1970s were not at all a problem in academia and Critical Theory was very much under the influence of neo-marxist political and social thinking. At university in the early 1970s, there were hard-liners, but one was mostly exposed to Marxism-Light as I would call it. With this I mean, the analyses of socio-economic powers and dynamics, taken out of the Marxist view of inevitable collapse of capitalism and then post-capitalist utopia of the salvation state which had already then not proved realistic in rapidly changing and adapting capitalist economies. However, when re-reading the two collections edited by Rose and Rose from 1976, which I read in 1977, that provide a series of articles on science and society, from an downright Marxist perspective, the Marxist jargon, the mentioning of the blessings of Maoism and the illusion of the end of capitalism and the bourgeoisie is quite weird. Indeed, Stalinism and Leninism and then the Cold War as discussed had blocked these analyses of science and society in the US. Ravetz was most of his professional life affiliated with the Centre for Philosophy and History of Science in Leeds where he worked for a short period of time with Toulmin. Ravetz in his book presents a comprehensive analysis of science and research, starting with problems that he expected would become more prominent. He discusses in depth the consequences of what he called 'the industrialization' of science which goes against the Mertonion norms with its protection of property and top-down management. He argued that because of enormous increase in scale, loss of social and ethical control, the system would increasingly face poor quality 'shoddy' research because of the lack of shared value of individual researchers with the scientific community. On the other hand, he is deeply concerned about the external influences on the research agenda by powerful private parties, multi-nationals, but also the military and governments. We

(continued)

know noe that Ravetz writing that book at that time was quite visionary. During his whole career he studied issues of uncertainty, risks and unwanted effects attached to the use of novel scientific knowledge and technology in society (Funtowicz & Ravetz, 1990; Ravetz, 2011). He wrote about the ethics of science and scientists and criticizes the claim 'of neutrality' that was used by researchers to evade their social responsibility. At that early stage preparing for my professional life, reading this book for me was truly a transformative experience and Jerry Ravetz was an inspiration and it was special that he participated when in the late fall of 2012 through 2013 we prepared for the start of *Science in Transition* described in Chap. 3.

The huge impact of Thomas Kuhn's *Structure of Scientific Revolutions,* published in 1962, has already been mentioned many times. It has opened up the debates in the history and sociology of science, but at the same time affected the domain of the philosophers showing through historical and sociological research the problems of logical positivism. Kuhn presented a descriptive account of what scientists do, which sociologically, but also (methodo)logically deviates from the normative positivist scientific method. He did however not provide judgement about the way science was actually done from the philosophical perspective (positivism) and did not propose an alternative correct formal method. This, in the eyes of his critics, was not logic or if it was logical, they did not agree. They asked the question whether Kuhn's description wasn't in fact normative. They make it, Kuhn writes in discussion with his critics, clear that they don't like his normative prescriptions using terms as *'corrupt our understanding and diminish our pleasure'* and *'a plea for hedonism'*. (Lakatos & Musgrave, 1970) They accuse Kuhn not using logic while they themselves use normative non-cognitive arguments and language. (p237). *'History and social-psychology are not, my critics claim, a proper basis for philosophical conclusions'* (p235). This is an important issue as it points to the gap between the philosophy and the practice of science. *Criticism and the Growth of Knowledge, eds. Lakatos and Musgrave* (Lakatos & Musgrave, 1970) is based on the contributions to a symposium held 13 July 1965 in London. In the final chapter, *Reflection on my Critics,* Kuhn declares his epistemological viewpoints that are beyond positivism, foundationalism and Popper's theory of falsification, but not sceptic nor relativistic. In fact, Kuhn states that his descriptive account of the process of inquiry at the same time indeed is normative. Because, if you want your inquiry to succeed you should use that process, that scientific method, which of course involves logic, mathematics, statistics and other accepted methods at a given moment in time in a research community. Indeed, science as Kuhn concluded, is a process of the community and not of an individual. A lot of the discussion in *Criticism* in my reading then indeed was about the differences between the descriptive historical and in some respect sociological mode of Kuhn's approach versus the normative mode of especially Popper and to some degree Lakatos. Popper admits that normal science exists, but finds it degrading and compares it to applied science and warned for the dangers normal science could pose to science. This is very reminiscent of the elitist scientific attitudes Snow and Medawar were criticising. Popper even suggested Kuhn did not seem to dislike normal science, whereby he exhibited his normative way, not only of theorizing about science, but also of judging scientists (p52, 53).

I cite some of the most interesting lines of Kuhn:

'I am no less concerned with rational reconstruction, with the discovery of essentials, than are philosophers of science. My objective, too, is an understanding of science, of the reasons for its special efficacy, of the cognitive status of its theories. But, unlike most philosophers of science, I began as an historian of science, examining closely the facts of scientific life'

Kuhn *'discovered that much scientific behaviour, including that of the very greatest scientists, persistently violated accepted methodological canons,...' p236*

In the current context of course the question is: who exactly had accepted these canons? Philosophers, but apparently not researchers! In response to Lakatos, Kuhn describes succinctly his conceptual frame:

'some of the principles deployed in my explanation of science are irreducibly sociological, at least at this time. In particular, confronted with the problem of theory-choice, the structure of my response runs roughly as follows: take a group of the best available people with the most appropriate motivation; train them in some science and in the specialties relevant to the choice at hand; imbue them with the value system, the ideology, current in their discipline (and to a great extent in other scientific fields as well); and, <u>finally, let them make the choice</u>. If that technique does not account for scientific development as we know it, then no other will. There can be no set of rules of choice to dictate desired individual behaviour in the concrete cases that scientists will meet in the course of their careers. Whatever scientific progress may be, we must account for it by examining the nature of the scientific group, discovering what it values, what it tolerates, and what it disdains. That position is intrinsically sociological, and, as such, a major retreat from the canons of explanation licensed by traditions which Lakatos labels justificationism and falsificationism, both dogmatic, and naïve'. p237, 238.

It is important to take note that Lakatos, in his one-hundred-page long contribution to this book, wrote that this debate *'did not start with Kuhn. An earlier wave of 'psychologism' followed the breakdown of justificationism. For many, justificationism represented the only possible form of rationality: the end of justificationism meant the end of rationality ....After the collapse of Newtonian physics, Popper elaborated new, non-justificationist critical standards. Finding them untenable, they identify the collapse of Popper's naïve falsificationism with the end of rationality self.'* P178. Lakatos, a true Popperian and believer in the 'scientific method' at that time, started to work on his concept of Research Programmes, a mix of Popperian and Kuhnian thought.

In the remainder of the chapter, Kuhn responds to the critique that his description of science, without a rejection of the methods used, opens the doors to relativity and nihilism. It was argued that personal opinion, mob psychology and elites with power and vested professional interests could determine the outcome of discussions regarding theory choice. He cites the non-cognitive, but important criteria and values that are being used and accepted in communities of inquirers and have been implied by Popper in his normative description of theory choice, including '*accuracy, scope, simplicity, fruitfulness*'. (P261, 262) Kuhn emphasizes that these are not rules that can be applied in a straightforward manner and his historical research has shown that they may evolve and change over time in the community.

When Kuhn prepared his book, in the late 1950s, logical positivism despite the prominent works by Quine and Sellars, still ruled in the philosophy of science and pragmatism was not considered to be a sound and fruitful alternative. Many still believed that the problems of positivism could be solved by analytical philosophy. But Kuhn's analyses and conclusions as expressed above are, although not cited by him, reminiscent of American pragmatism and the critiques of Peirce and Dewey on the dominant philosophy of science of their times.

John Ziman (1925–2005) was a physicist who between 1960 and 2000 was one of the first to write systematically, in depth and broadly about science. In 1968 he published *Public Knowledge* (Ziman, 1968) his first of nine books on science and as Jerome Ravetz, Ziman's contemporary colleague and science writer, in his obituary wrote: '*In this he bypassed the debates among the philosophers who saw science as a collection of "theories" requiring some sort of logical proof; for him the essential feature of scientific knowledge is its social character.*' (Jerry Ravetz, Guardian, February 2005).

It's Anthropology, Stupid!

In most of their books, Toulmin, Hanson, Ravetz and Ziman, but also Polanyi, take all aspects of the scientific enterprise into account in their analyses of how consensus regarding reliable knowledge is produced and thus what distinguishes science as a social activity. In their opinion it is exactly the complex of the methods, personal psychology, the community and the sociology of the researchers in organizations that determines what science is. Their writings went against the widely held believes about science and as a consequence were virtually neglected by main-stream philosophy, history and sociology. Because of its multidisciplinarity, in addition their work did not belong to one of these classical academic disciplines. Similarly, even Bruno Latour in his *We were never Modern* complained about the slow recognition of Latour and Woolgar's *Laboratory Life* by philosophers and sociologist of science. (Latour, 1993; Latour & Woolgar, 1979) This was duly confirmed by Hacking in his very late 1988 (!) review of '*Laboratory Life*'. With regard to this seminal book of 1979 and his own 1983 book that I here cite a lot and wherein he argues to take a look at the practice of science, he declares that '*it was shameful not to examine the one outstanding piece of work then available that took laboratory science seriously and argued the strong anti-realist doctrine in existence*'.(p278) (Hacking, 1988) Latour pointed out that we do accept anthropology crossing all these academic territories, but apparently we do not allow this for an anthropology of the tribe of humanity that is involved in science.

This may, Ravetz believed, be the reason that this type of work has had relatively little impact (Ravetz, personal communication 2013). That may well be the case, but as argued above, meta-science research drew in general very little attention from those active in research in the academic disciplines or in the 'corridors of power' of academia (Miedema, 2012). Lack of impact has also been blamed on the fact that the work of these authors lacked a novel theory, theoretical frame or a specific novel concept. Exceptions to this are Polanyi's concept of tacit knowledge and Toulmin's metaphor of maps for theories (1953) and his evolutionary concept of progress in science (1972). I disagree with this critique, as I my opinion the main hypothesis for which they provided evidence and which is the basis for this book, is that in the history of science, the dominant image of science which proved philosophically wrong around 1960, was strongly politically and culturally determined and has until now distorted and hurt the practice of science in many different ways. It is on the basis of these insights, that many scholars have since then began to study the practice of science. These studies in the recent past have resulted in renewed movements to improve the practice of science and make it more suitable to contribute to solve the grand challenges of the twenty-first century. John Ziman already in his early books *Public Knowledge* and *Reliable Knowledge* has provided insights from the trenches of science about the

(continued)

problem of the myth of the 'scientific method', which at that time still few others understood and for which Ziman later coined the term 'the Legend' (Ziman, 2000). He wondered in 1968 (!) (Ziman, 1968) how this *'logico-inductive' metaphysics of Science.. can be correct, when few scientists are interested in (it) or understand it, and no one ever uses it explicitly in his work? But if Science is not distinguished from other intellectual disciplines neither by a particular style or argument nor by a definable subject matter, what is it?* (p8). He then sketches the social process of inquiry, hypothesis, testing and criticism and states that *'it is not a subsidiary consequence of the 'Scientific Method'; it is the scientific method itself.' 'The defect of the conventional philosophical approach to Science is that it considers only two terms in the equation. The scientist is seen as an individual, pursuing a somewhat one-sided dialogue with taciturn Nature. But it is not like that all. The scientific enterprise is corporate. It is never one individual that goes through all steps of the logico-inductive chain; it is a group of individuals, dividing their labour but continuously and jealously checking each other's contributions'.* (p9)

John Ziman could in those days, find virtually no literature on consensus building by the community and the social process and *'that makes the Philosophy of Science nowadays so arid and repulsive. To read the latest volume on this topic is to be reminded of the Talmud...' It is fiercely professional and technical and almost meaningless to the ordinary working scientist. This is unfortunate ..I shall try to heal the breach by talking semi-philosophically about the intellectual procedures of scientific investigation.'* (p31).

In *Reliable Knowledge: an exploration of the grounds for belief in science (J. M.* Ziman, 1978*)* an important book in this context, Ziman did the same bypass as in *Public Knowledge* as in all his books regarding the philosophical basis of the Legend. In the introductory paragraph 1.4 he firmly states that from data, diagrams, models or pictures, *'meaning cannot be deduced by formal mathematical or logical manipulation. For this reason scientific knowledge is not so much 'objective' as 'intersubjective' and can only be validated and translated into action by intervention of human minds'* (p7). Ziman is very realistic and knows the daily practice of physics and does not conceal weaknesses known to investigators but disguised by the believers of the Legend: *'The achievements of intersubjective agreement is seldom logically rigorous; there is a natural psychological tendency for each individual to go along with the crowd, and to cling to a preciously successful paradigm in the face of contrary evidence. Scientific knowledge thus contains many fallacies, mistaken beliefs that are held and maintained collectively and which can only be dislodged by strong persuasive events.'* (p8.) He describes how scientist are 'brainwashed' during their training in the concept, accepted beliefs and methods in the current paradigms of their field. He explains in great detail and nuance how in the 'social model of science', the scientific community produces the knowledge we designate as scientific knowledge and what makes it unique and reliable. Ziman builds further on the work of those who criticised positivism and the Legend - Polanyi, Hanson, Toulmin and Kuhn- published in the decade before. Ziman points to fact that there is not one scientific method, there are many dimensions to scientific knowledge and *'that explains the strange sense of unreality that scientist feel when they read books about the philosophy of science' (p84)*. From this point of view from the natural sciences, he concludes that the social sciences and humanities of course can produce reliable scientific knowledge and he states in an unexpected humanistic lyric paragraph *that 'the challenge to the behavioural sciences is not coming from physics but from the humanities'.* (p185).

Jerome Ravetz in his, in the STS fields well-known, *Scientific Knowledge and its Social Problems* presented a unique philosophical-sociological analysis. (Ravetz, 1971) It provides an integrated very rich view of science, its theoretical assumptions, its ideologies, power games, issues of ethics and social responsibilities and the sociology and politics of the system and the interaction with society. Ravetz cites a broad body of the most relevant scholars at that time. He refers frequently to the work of his temporaries Toulmin, Ziman, Rose and Rose and especially Polanyi's *'Personal Knowledge'* (Polanyi, 1958, 1962). He really 'took a look' at the practice of science and especially emphasizes science as craftmanship and subsequently discusses the philosophical assumptions about the special status of theories and how knowledge is produced. He, on the basis of his understanding how science and research is being done, rejects the positivist and foundationalist ideas With respect to 'the scientific method' and positivism he clearly states that in research the underlying *'principles and precepts that are social in their origin and transmission, without which no scientific work can be done.. guide and control the work of scientific inquiry.'* (p146) More explicitly: *'The individual scientist; and the criteria of adequacy are set by his scientific community, not by Nature itself.'* (p149) With respect to *'the maturity of a field an important part lies in the strengthening of the criteria of adequacy. This is not all of course; the development of new tools, and the creation of an appropriate social environment are equally important. Nor can the strengthening of criteria of adequacy be done in an abstract, automatic fashion, as by attempted imitation of a succesful field* (p157). About the relation between philosophy and the practice of science he says: *'Philosophers of science have attempted, with some success to provide a rationale for the different basic patterns of argument, showing why it is reasonable for an intelligent person to place reliance on them....But as these philosophical arguments become more refined and sophisticated, they drift further and further from the practice of science.'* Finally, for the present discussion it is of interest to close with the following citation on the dichotomy of values and facts. Ravetz, unlike Polanyi, but like Bernal whom he also personally knew, sees research primary as a social activity that needs conscious strategies to be able to make proper judgements regarding problem choice. He explicitly mentioned values other than strict cognitive arguments that have to enter into these evaluations (p161). *'The criteria of value, and judgements based upon them, form an interesting contrast to those of adequacy. ...we shall find ourselves involved in problems of the social activity of science. ...The exclusion of problems of value from the traditional philosophy of science has its roots in the ideology of modern natural science as it was formed through many generations of struggles.....the considerations of social value by which all other human activities are assessed were declared irrelevant'* p160.

Mary Hesse (1924–2016) studied mathematics, physics and philosophy and taught mathematics and philosophy at several universities in England. She has written extensively on the philosophy of science. Mary Hesse wrote in 1972: *'During the last half-century much of professional Anglo-American philosophy of science has been devoted to detailed development of internal logic of natural science based on empiricist criteria, and also on attempts to show how this logic applies also in the social sciences and in the study of history. Suggestions....to the effect that there are other modes of knowledge than the empiricist were sometimes actively resisted but more usually totally disregarded'.* *'It was held that adoption of at least a modification this empiricist method is required for human sciences* *'to attain knowledge status at all' which in her view is* *'imperialism claimed for natural science'* (p27) (Hesse, 1972).

'These distinctions that I believe are made largely untenable by recent more accurate analyses of natural science.

1. *In natural science experience is taken to be objective, testable, and independent of theoretical explanation. In human science data are not detachable from theory, for what count as data are determined in the light of some theoretical interpretation, and facts themselves have to be reconstructed in the light of interpretation.*
2. *In natural science theories are artificial constructions or models, yielding explanation in the sense of logic of hypothetic-deduction: if external nature were of such a kind, then data and experience would be as we find them. In human science theories are mimetic reconstructions of the facts themselves, and the criterion of a good theory is understanding of meanings and intentions rather than deductive explanation.*
3. *In natural science the law-like relations asserted of experience and external, both to the objects connected and to the investigator, since they are merely correlational. In human science relations asserted are internal, both because the objects studied are essentially constituted by their interrelations with one another, and also because the relations are mental, in the sense of being created by human categories of understanding recognized (or imposed? By the investigator.*
4. *The language of natural science is exact, formalizable, and literal; therefore, meanings are univocal, and a problem of meaning arises only in the application of universal categories to particulars. The language in human sciences is irreducibly equivocal and continually adapts itself to particulars.*
5. *Meanings in natural science are separate from facts. Meanings in human science are what constitute facts, for data consist of documents, inscriptions, intentional behaviour, social rules, human artefacts, and like, and these are inseparable from their meanings for agents.*

'Let us however concentrate for a moment on the natural science half of the dichotomy what is immediately striking about it to readers versed in recent literature in philosophy of science is that almost every point made about the human sciences has recently been made about the natural sciences. And that the five points made about the natural sciences presuppose a traditional empiricist view of the natural science that is almost universally discredited' (p277) (Hesse, 1972)

Richard Rorty's *Philosophy and the Mirror of Nature* published in 1979 had enormous and immediate impact and for most scholars of pragmatism was the start of the pragmatic turn (Rorty, 1979). Rorty, in chapters III and IV, starts by discussing in depth the serious critiques of Quine and Sellars on the classical dichotomies of logical positivism. In addition, he took the pragmatic turn in chapter VII discussing at length Kuhn's work and putting it firmly in the larger context of the pragmatism of John Dewey. He concludes that 'analytic' epistemology (i.e. "philosophy of science") became increasingly historicist and decreasingly "logical" (as in Hanson, Kuhn, Harré and Hesse) (p168). He discusses the 'behavioristic' critiques of Quine and Sellars, following Wittgenstein's *Philosophical Investigations* published at the same time in 1953, on *'the two distinctions the "given" and "that what is added by the mind" and that between the "contingent" (because influenced by what is given) and the "necessary" because entirely "within" the mind and under its control)...he presents them as forms of holism. As long as knowledge is conceived of as accurate representing- as Mirror of Nature- Quine's and Sellar's holistic doctrines sound pointlessly paradoxical, because such accuracy requires a theory of privileged representations, ones which are automatically and intrinsically accurate. ...I shall be arguing that their holism is a product of their commitment to the thesis that justification is not a matter of a special relation between ideas (or words) but of conversation, of social practice. ...we understand knowledge when we understand the social justification of belief, and thus have no need to view it as accuracy of representation.'(p170)...*this is, Rorty says, *'the essence of what I shall call epistemological behaviorism, an attitude common to Dewey and Wittgenstein'. (p174) 'Epistemological behaviorism (which might be called "pragmatism" were this term not a bit overladen)...is the claim that philosophy will have no more to offer than common sense (supplemented by biology, history, etc) about knowledge and truth. (p176).* The term 'behavioristic' may seem peculiar, but refers to the social process by which a community of inquirers come to produce and accept knowledge and beliefs.

In the pages that follow Rorty dispenses with foundationalism and even with philosophy at large, the latter goes much too far for philosophers like Kitcher, who see enough problems to philosophize about. Indeed, since the demise of the Legend, there is no systematic 'grand unified theory' in the philosophy of knowledge. As I will argue in Chap. 4, pragmatism has a lot to offer with regard to our understanding and philosophizing about knowledge and knowledge production. As Rorty discussed (p367), it may not provide a systematic alternative, but it does provide a hermeneutical method and viewpoint about science and inquiry (see also Kuhn The essential tension p xiii and xv). This, to many a philosopher of the analytic tradition may have been disappointing and the main reason to not take pragmatism serious as philosophy, but must be understood in that pragmatism is a reaction by 'peripheral' philosophers (James, Dewey, Wittgenstein, Heidegger) to a 'systematic' philosophy which Rorty designates a mainstream analytic 'superstition'. These 'peripheral' philosophers are according to Rorty the 'edifying' philosophers. They do not provide a system with a set of rules but offer <u>moral and intellectual instructions and enlightenment.</u>

As Flyvberg (2001) argues, hermeneutics is not only relevant for the social sciences but also for the natural sciences *'as it is now argued that natural sciences are historically conditioned and require hermeneutic interpretation. Natural scientist, too, must determine what constitutes relevant facts, methods, and theories; for example, what would count as "nature".* (p28).

Nancy Cartwright, a mathematician and philosopher who has studied the practice of physic in relation to the myths of analytical philosophy. She wrote *The Dappled World* (Cartwright, 1999) in follow up of *How the Laws of Physics Lie* (Cartwright, 1983), in which she discusses the classical ideas of the unity of science and the myth of the universality of physics and she takes for comparison economics, the discipline that is famous for imitating (or since the financial crisis having imitated?) physics. The physics that never was, as Cartwright shows. *The Dappled World* is a very technical book, but its conclusions (p9 and 10) are clear theories and claims have been stablished in very artificial settings in the laboratory or as in economics by keeping everything else the same (*ceteris paribus*) both which in the real world are rare to occur: *'I conclude that even our best theories are severely limited in their scope. For, to all appearances, not many of the situations that occur naturally in our world fall under the concepts of these theories.....' 'The logic of the realist's claim is two-edged: if it is the impressive empirical successes of our premier scientific theories that are supposed to argue for their 'truth'...then it is the theories as used to generate these empirical successes that we are justified in endorsing. How do we use theory to understand and manipulate concrete things- to model particular physical or socio-economic systems? The core idea is ... the belief in one great scientific system, a system of a small set of well-co-ordinated first principles admitting a simple and elegant formulation, from which everything that occurs, or everything of a certain type or in a certain category that occurs, can be derived. But treatments of real systems are not deductive, ...(not) even if we tailor our systems as much as possible to fit our theories, which is what we do when we want to get the best predictions possible.'*

This is the reason, and that is well known, why many drugs shown to have beneficial effects in a highly selected patient population and well-controlled clinical trials, don't do as well in clinical practice. Cartwright got a lot of criticism to the kind of criticism she articulated in *How the Laws of Physics Lie* but her response is clear, and relates to the myth of the Legend: *'I agree that my illustrations ....are 'a far cry' from showing that the system must be a great scientific lie. But I think we must approach natural science with at least as much of the scientific attitude as natural religion demands'.*

Her examples are from physics, economics, medicine and genetics. Her conclusions reminds on the one hand of the arguments of Nagel discussed above, and on the other hand of the persuasive work of Richard Lewontin, which in a less analytic and technical way, criticizing the ideologies of biology, genetics, molecular biology and the dream of the human genome project and thus of the positivist molecular-biologists and clinicians-researchers who believed would reductionist science solve the problem of our diseases- cancer, cardiovascular, and mental illnesses alike. (Lewontin, 2000; Lewontin et al., 1984).

Hillary Putnam (1926–2016) was a mathematician and philosopher who has had a broad and deep impact on mathematics, ethics and the philosophy of science. He is famous and admired for his critical thinking about the work of others, and interestingly, as well as about his own work and has as consequence changed his philosophical ideas and positions several times in his long career. He started as a student with Hans Reichenbach, a major figure in pre-war analytical philosophy. Via positions amongst others at Princeton and MIT he worked at Harvard until 2000. In his later years he wrote widely about American pragmatism (Putnam, 1995; Putnam & Conant, 1994) and in particular how it could overcome the problems of the analytical philosophical tradition including foundationalism, and the various dualisms such as the analytic-synthetic, the objective-subjective and the fact-value dichotomies. His *Reason, Truth and History* (Putnam, 1981) is illuminating with respect to the flaws of the positivist philosophy of the Legend. In particular Chap. 3, but also more broadly the thinking presented in Chap. 8 are insightful. In 2004 he published *The collapse of the Fact/Value dichotomy* (Putnam, 2002) where he discusses how most 'analytical philosophy of language and much metaphysics and epistemology has been openly hostile to talk of human flourishing, regarding such talk as hopelessly "subjective"- often relegating all of ethics, in fact, to that waste baker category' (p viii), and he argues for the economics approach of Amartya Sen. He delves deep, as always, and I will leave that to the more experienced reader but here I cite the very last paragraph which is in plain English but boldly worded which makes his position after a lifetime hard work on exactly these matters very clear:

'I have argued that even when the judgments of reasonableness are left tacit, such judgments are presupposed by scientific inquiry (indeed, judgments of coherence are essential even at the observational level: we have to decide which observations to trust, which scientists to trust-sometimes even which of our memories to trust.) I have argued that judgments of reasonableness can be objective, and I have argued that they have all of the typical properties of value judgments. In short, I have argued that my pragmatist teachers were right: "knowledge of facts presupposes knowledge of values." But the history of the philosophy of science in the last half century has largely been a history of attempts - some of which would be amusing, if the suspicion of the very idea of justifying a value judgment that underlies them were not so serious in its implications- to evade this issue. Apparently any fantasy -the fantasy of doing science using only deductive logic (Popper), the fantasy of vindicating induction deductively (Reichenbach), the fantasy of reducing science to a simple sampling algorithm (Carnap), the fantasy of selecting theories given a mysteriously available set of "true observation conditionals," or, alternatively "settling for psychology" (both Quine)- is regarded as preferable to rethinking the whole dogma (the last dogma of empiricism?) that facts are objective and values are subjective and "never the twain shall meet." That rethinking is what pragmatists have been calling for for over a century When will we stop evading the issue ("knowledge of facts presupposes knowledge of values.") (insert FM) *and give the pragmatist challenge the serious attention it deserves?* (p145)

I have in this philosophical time-travelling now arrived in the twenty-first century. I want to discuss Philip Kitcher's work, which for several reasons is of interest in this context. Starting like Putnam from the analytical science tradition, he has described his intellectual history since the 1980s, in the beginning criticizing some and defending other parts of the Legend but gradually losing faith. Kitcher has been reflecting on the philosophical transition he went through, from empirical positivism, natural empirism to a form of neopragmatism. Even in times when the more general pragmatic turn was already going on in the field (Bernstein, 2010; Putnam & Conant, 1990), he experienced how different this philosophical approach was, not in the least in the eyes of his mainstream analytically thinking peers (Kitcher, 2012). Kitcher in 1999 was appointed as John Dewey Professor of philosophy at Columbia. From his website: '*Following Dewey, I believe in the need for a reconstruction of philosophy (so that it will not be a "sentimental indulgence for the few"), and I worry about the increasing narrowness and professionalization of academic philosophy. In working with graduate students, I hope to instil a capacity for clarity and rigor without sacrificing the sense of why philosophy matters.*'

In his *The Advancement of Science* (Kitcher, 1993), which carries the strong subtitle "*Science without a Legend, Objectivity without Illusion*', this struggle is throughout the book most visible, but Kitcher is to be recommended for being very explicit about it upfront and in the epilogue: '*Once, in those dear dead days, almost, but not quite, beyond recall, there was a view of science that commanded wide spread popular and academic assent*'…. '*Legend celebrates scientists as well as science*'. ......*scientists have achieved so much through the use of the SCIENTIFIC METHOD.*'..'*there are objective canons of evaluation of scientific claims; by and large, scientists (at least since the seventeenth century) have been tacitly aware of these canons and have applied them in assessing novel or controversial ideas....*'(p3).

'*So much for the dear dead days. Since the late 1950s the mists have begun to fall. Legend's lustre is dimmed. While it may continue to figure in textbooks and journalistic expositions, numerous intelligent critics now view Legend as a smug, uninformed, unhistorical, and analytically shallow. Some of the critiques, science bashers, regard the failure of science to live up to Legend's advertising as reason enough to question the hegemony of science in contemporary society. I shall not be concerned with them, but with the critiques of the Legend bashers, those who believe that Legend offered an unreal image of a worthy enterprise.*'(p5) Kitcher acknowledges that although he believes that the classical philosophy '*belongs amongst the greatest accomplishments of philosophy of our century*', it has been shown to have its problems. He only once in a footnote (!) (p7) cites the devasting critique of Popper discussed above and admits that '*despite efforts of a few philosophers, little headway has been made in finding a successor for Legend. If anything, recent work in the history of science and the sociology of science has offered eve more sweeping versions of the original critiques*'......, *I am not **yet** ready to abandon the search for generality*' (bold applied by FM) p8.

Kitcher is much concerned with the objectivity of theory choice where indeed (social) criteria are at play which according to Legend are non-epistemic because external. He also wrestles many pages with the classical problem of representation of reality by theory and of realism of the objects of science and in these discussions uses, as per Legend, the success of natural science as kind of foundation, a warranty for objectivity and realism. This feels like causality reversed. Kitcher at that time believed that Legend could philosophically and sociologically be rescued, in his way or another. He believed that *'the Legend was broadly right about the characteristics of science. Flawed people, working in complex social environments, moved by all kind of interests, have collectively achieved a vision of parts of nature that is broadly progressive and rests on arguments meeting standards that have been refined and improved over centuries. Legend does not require burial but metamorphosis.' p390* This defence of Legend is remarkable since writing this in 1993, he is aware and discusses the seminal work of the scholars who convincingly showed, as I discussed above, that the myth of 'the scientific method' and its normative canons, never did relate much to daily practice of inquiry and the idea of foundationalism did not hold. Kitcher (p10) admits that the Legend was a normative construction, but incorrectly seems to suggest it came from studying science and can be rescued by studying the practice of science again. Kitcher was at that time critized by Shapin (cited by Kitcher p303) that he still worked from the Legend's 'individualism' of the scientist instead taking the work of many scholars to heart that shows the social process and the community of inquiry in practice. Very interestingly, in the final pages he suggests that philosophy should be normative and could suggest ethics and values for how the enterprise of science could (and should) be organized to optimally contribute to human flourishing: *'Yet even if the metamorphosis of Legend attempted here clears away those errors, it does not address the issue of the value of science. To claim as I have done that that the sciences achieve certain epistemic goals that we rightly prize is not enough- for the practice of science might be disadvantageous to human well-being in more direct was, practical ways. A convincing account of practical progress will depend ultimately on articulating an ideal of human flourishing against which we can appraise various strategies for doing science. Given an ideal of human flourishing, how should we pursue our collective investigation of nature........how should we modify the institution as to enhance human well-being?.... The philosophers have (no the Legend has .., FM) ignored the social context of science. The point however is to change it."* (p391) I will return to the later work of Kitcher, which shows his sharp pragmatic turn, when this topic is discussed further in Chap. 4.

Helen Longino (born 1944) has focussed throughout her career as philosopher on the social character of scientific inquiry. She is motivated in this work by Women's Studies, the role of social values and criteria, equality, gender and inclusiveness. She has studied it from different theoretical and practical viewpoints. She understands the Legend and the struggle of the classical philosophers, including Kitcher, to break free from the classical view of the scientific method, the Legend. She is avoiding the extreme, that there is no objectivity in scientific inquiry at all, argued by those who claim that it is determined by values and interests only and unconstrained by empirical observations. In her widely appreciated *'Science and Social Knowledge'*(Longino, 1990) her tour the force on this is described for the first time in an analysis contrasting the logical positivist

philosophy of Hempel with the 'Wholism', as she calles it, of Hanson, Kuhn and
Feyerabend. She goes basically through the same intellectual moves as the writers
cited above and, in the end, tries to present a contextual empiricist 'scientific
method' that is truly social in which the community of inquirers also takes social
values pertinent to the context of the work into account. *'My concern is that with a
scientific practice perceived as having true or representative accounts of its subject
matter as a primary goal or good. When we are troubled about the role of contextual
values or value-laden assumptions in science, it is because we are thinking of scientific inquiry as an activity whose intended outcome is the accurate understanding of
whatever structures and processes are being investigated. If that understanding is
itself conditioned by ours or others' values, it cannot serve as a neutral and independent guide.'* Against this she argues: *'The dichotomy of these approaches should
not be seen so much as a contraction to be resolved in favour of one or the other
position, so much as reflective of a tension within science itself between its
knowledge-extending mission* (application in contexts) *and its critical mission* (better theories)' p34.

> *'In assessing particular research programmes, it is important to keep in mind that knowl-
> edge extension* (testing the effects of claims in experimental and real-world settings) *and
> truth* (as accepted beliefs, Longino must mean to say) *can guide scientific inquiry and
> serves as fundamental, but not necessarily compatible, values determining its assessment.'
> *Thus, while a demonstration of the contextual value ladeness of a particular research pro-
> gram may serve to disqualify it as a source of unvarnished truth about its subject matter,
> such demonstration may have little bearing on one's assessment of it as an example of sci-
> entific inquiry.'(p36)* (non-italic inserts are mine).

There is in Longino's method, her epistemology, no timeless foundation, but
there are background assumptions, ethical, political, social and other, and there is a
practice of reasoning about them. They are under scrutiny, with full criticism and
eventual acceptance by the community of inquirers thus correcting for subjective
individual preferences (p216). These assumptions, like the classical scientific methods, are not insensitive to cultural and political changes brought about over periods
of time by changes in the world views of citizens wherever they live their life. The
myth or the Legend, Longino correctly observed, has served as a timeless and sta-
bile disguise providing an account that can *'render invisible the background assumptions. The methodologies associated with logical positivism did render them
invisible, which is, I suspect, one reason they remain persuasive among scientists
even after being abandoned by philosophers......The myth of value neutrality, that
is the consequence of the more general view that scientific inquiry is independent of
its social context, is thus a functional myth.' (p225).*

This is an important insight. In fact, by employing this myth of neutrality, scientific inquiry and science as a knowledge system in society is in first instance mainly
conservative, resisting critique regarding its accepted theoretical core, and its reflection on its own societal activity. It prohibits, or at least discourages on methodological (epistemic) grounds, also the critique through scientific inquiry of the institutions
and social developments and conceals the interaction of science with public and
private power structures in society. This is as an example reflected in the negative

response of Polanyi and Russell, key opinion leaders in UK physics on BBC radio broadcasted the beginning of 1945, to a caller's question if something of practical use could be expected to be done with quantum physics. Much later in 1962 Polanyi *'actually,"* admits, *"the technical application of relativity...was to be revealed within a few months by the explosion of the first atomic bomb."* 'Polanyi *argued that because science is unpredictable, then its subsequent technical and social outcomes are even more so. He weaves an intricate analogy between the conduct of science and the play of the economic market, both of which exemplify how individuals can maximize socially beneficial outcomes by pursuing their own interests and adjusting, mutually but independently, to the interests of others. The same "invisible hand" that guides the market guides science. While he allows that "Russell and I should have done better in foreseeing these applications of relativity in January 1945," he extends their own incapacity back a half century by also arguing that "Einstein could not possibly take these future consequences into account when he started on the problem which led to the discovery of relativity" because "another dozen or more discoveries had yet to be made before relativity could be combined with them to yield the technical progress which opened the atomic age"* (Cited in (Guston, 2012) Guston 2012 Minerva). A bit dubious this evasion of one of the major ethical and political issues of twentieth century science, since Einstein and Szilard having fled the Nazi's to the US, in 1939 urged Roosevelt to get an atomic bomb build before Hitler did. Its deployment against Japan had not been the idea of a pacifist Einstein and many involved scientists, they instead had seen it as a major means of deterrent. Einstein was until his dead active in the Federation of Atomic Scientists and the Pughwash Conferences against proliferation of nuclear arms.

Longino concludes that this myth of neutrality is detrimental to major aspects of the practice of modern science in chapters on research on sex differences, and the genetics and biology of behaviour where 'hard' data is interpreted based on uncontested hidden social assumptions. Inquiry explicitly investigation and criticizing these cultural assumptions is per Legend declared non-scientific though, because of contextual assumptions that are made explicit.

Ten years after, in *'The Fate of Knowledge'* (Longino, 2002) she has gone further down the road, further away from the timeless certainty of the Legend. She writes: *'My aim in this book is the development of an account of scientific knowledge that is responsive to the normative uses of the word "knowledge" and to the social conditions in which scientific knowledge is produced. Recent work in history, philosophy, and social and cultural studies of science has emphasized one or the other. As a consequence, accounts intended to explicate the normative dimensions of our concept- that is elaborating the relation of knowledge to concepts such as truth and falsity, opinion, reason, and justification- have failed to get a purchase on actual science, whereas accounts detailing actual episodes of scientific inquiry have suggested that our ordinary normative concepts have no relevance to science or that science fails the test of good epistemic practice. That can't be right. The chapters that follow offer a diagnosis of this stalemate and an alternative account. I argue that the stalemate is produced by an acceptance by both parties of a dichotomous understanding of the rational and the social.'* (p1).

This is one of the main problems in science and academia, nearly 20 years later because we still see this stalemate and in our debate about science its characteristic discourse. Longino addresses the underlying assumption of this classical dualism of the Legend and rejects them, which opens up the possibility of a concept of science where internal and external criteria of value both can be used to make choices in science. She in 2002 immediately (on p3) goes to the work of Mill, Peirce and Popper who early on realised that science and the method used to come to accepted beliefs is not an individual but a truly social process, which as we have discussed goes against the Legend. Regarding Popper she points out correctly that Popper, as cited above, praised philosophers who involve in their analyses *'theories, and procedures, and, most important, (of) scientific discussions'*, *'contingent factors operating in the world of human affairs are beyond his epistemology'*. *'Unlike discussions by Mill and Peirce, Popper's theory of knowledge deliberately bypasses the connection to science and inquiry as practiced and remains the ideal'* (p7). I cite her own resume of the book which is mainly dealing with the problem of what she calls the Rational-Social Dichotomy which as we saw is a main pillar of the Legend: *'The work in social and cultural studies has stimulated a range of responses from philosophers. Some simply rejected the relevance of this work to philosophical concerns, or ....have seen it as empirically and conceptually misguided. Some like Philip Kitcher...have tried to take the sting out of it, by sifting through the claims of the sociologists and sociologically oriented historians attempting refutation of those they deem extremist, and then incorporating a sensitivity to history or sociological analysis into their constructivist accounts of inquiry. ..., I argue that these efforts, too, are vitiated* by *a commitment to the dichotomy of rational and the social. I offer an account of scientific knowledge that not only avoids the dichotomy but integrates the conceptual and normative concerns of philosophers with the descriptive work of the sociologists and historians.*

Longino aims to integrate in the understanding of scientific inquiry the fact that *'cognitive capacities are exercised socially, that is interactively'* and argues that more *'more complete epistemology for science must include norms that apply to practices of communities in addition to norms conceived as applying to practices of individuals. Following through on the consequences of the analyses breaking with conventional views of scientific knowledge as permanent, as ideally complete, and as unified and unifiable....means accepting provisionality, partiality, and plurality of scientific knowledge. ...I insist on an epistemology for living science, produced by real empirical subjects. This is an epistemology that accepts that scientific knowledge cannot be fully understood apart from its deployments in particular material, intellectual and social contexts.'* She makes it clear that there need to be pluralism in these epistemologies.

Longino wants to take advantage in her epistemologies of both the Rational and the Social and takes us through some technical chapters, in which she makes it clear that we have a lot to figure out if we (want to properly) use a mix of rational normative criteria from the philosophers and the social criteria and norms the sociologists have revealed. This is especially interesting knowing that scientists do use in their field validated standard, methods and accepted ways of reasoning, but do not take the normative canons of the Legend to seriously in their daily practice, whereas they use consciously and unconsciously the social norms and values derived from their cultural upbringing in all its aspects of a society. In the last ten pages she concludes that the Rational-Social classes of criteria and norms are thus not used in separation, '*sociality does not come into play at the limit of or instead of the cognitive. Instead, these social processes are cognitive. ....and the social epidemiologist must have resources for the correction of ..epistemically undermining possibilities.*' This is required since opening up to the social, the stakeholders in society, opens up to power games which may be to the disadvantage of those problems which are vulnerable to 'inappropriate exercise of authority and biases. This is as we discussed (in Chap. 1) a problem of all times, past and yet to come, because scientific inquiry is not autonomous, value-free and not neutral and is not guided by the invisible hand of the Legend who tells us how best to allocate our public and private funds. Longino offers at the very end of the book a set of questions that demonstrate that she sees a lot of problems here for philosophers to work on for instance how goal-oriented inquiry and '*different kinds of goal might affect philosophy and knowledge and practices.* She goes one step further and involves in these questions '*the institutional organizations and how they affect the content of knowledge*' and asks '*How can a society use science to address problems when scientific goals and community structures are not mutually aligned? These questions bring out the political dimensions of science and broaden our conception of what philosophy of science can be about.*'

Finally, she asks '*What kinds of institutional changes are necessary to sustain the credibility, and hence value, of scientific inquiry while maintaining democratic decision making regarding the cognitive and practical choices the sciences make possible and necessary? The fate of knowledge rest in our answers*'.

With these questions, that almost all philosophers of science like Popper consider '*beyond their epistemology* or *theory of knowledge and deliberately bypass*', we return to the main problems addressed here: how does the Legend still determine the ideas and politics of scientific inquirers which distorts the collective of scientific inquiry, causes the current problems of science. Legend and its legacy has detrimental effects on our interaction with society and their publics and thus the knowledge we produce, this is 'the fate of knowledge' Longino is concerned with. Longino, after her own struggle with the dualism of the Legend, boldly has been going where sociologists, physicists, chemists, historians, even anthropologists, but few philosophers have gone before. Still, the reviewers of the book who praised her for that, criticize her for not presenting a detailed epistemology. Longino knew how the work of Dewey and James had been received by the 'real analytical philosophers' of their times, that must have offered some consolation.

2.3 Conclusion

Towards a Realistic Pragmatist View of Science, Natural Science and Social Science and Humanities

From the late 1960s philosophers, sociologist and historians of science gradually, but definitely showed the Legend of the 'Scientific Method' to be untenable:

- **There is no one formal scientific method that leads us to the truth**
- **There is no God-given or timeless, universal foundation for such a method to build on**
- **Knowledge is arrived at, not by individuals in isolation 'talking to nature'**
- **There are many ways (methodologies) to do good research**
- **In sharing ideas and experimental results and methods, for debate and scrutiny in a rigorous and communitive process by the community of inquirers**
- **Inquiry is a social process producing reliable knowledge that produced objective (intersubjective) knowledge**
- **Research is guided by our common cognitive and cultural values, when tested in experiments and discussions with peers constrained by natural and social reality**
- **Knowledge is tested in interventions and (social) actions in practice**
- **It is then either rejected, improved or it is accepted for the time being**
- **Knowledge claims are fallible, absolute and always up to scrutiny and tests**
- **It is this communitive open, independent and transparent process that is unique to science which has produced knowledge which has been proven to be reliable over the past centuries.**

References

Barker, G., & Kitcher, P. (2013). *Philosophy of science: A new introduction.* Oxford University Press.

Ben-David, J., & Sullivan, T. A. (1975). Sociology of science. *Annual Review of Sociology, 1*(1), 203–222. https://doi.org/10.1146/annurev.so.01.080175.001223

Bernstein, R. J. (1983). *Beyond objectivism and relativism: Science, hermeneutics, and praxis.* Basil Blackwell.

Bernstein, R. J. (2010). *The pragmatic turn.* Polity.

Bourdieu, P. (1975). The specificity of the scientific field and the social conditions of the progress of reason. *Information (International Social Science Council), 14*(6), 19–47. https://doi.org/10.1177/053901847501400602

Bourdieu, P. (2004). *Science of science and reflexivity.* Polity.

Cartwright, N. (1983). *How the laws of physics lie.* Clarendon Press.

Cartwright, N. (1999). *The dappled world: A study of the boundaries of science.* Cambridge University Press.

Crewdson, J. (2002). *Science fictions: A scientific mystery, a massive coverup, and the dark legacy of Robert Gallo* (1st ed.). Little, Brown.

Descartes, R. (1968). *Discourse on method, and other writings*. Penguin.

Edmonds, D. (2020). *The murder of professor schlick: The rise and fall of the Vienna Circle*. Princeton University Press.

Fleck, L. (1979). *Genesis and development of a scientific fact*. University of Chicago Press.

Flyvbjerg, B. (2001). *Making social science matter: Why social inquiry fails and how it can count again*. Cambridge University Press.

Funtowicz, S. O., & Ravetz, J. R. (1990). *Uncertainty and quality in science for policy*. Kluwer Academic Publishers.

Geison, G. L. (1995). *The private science of Louis Pasteur*. Princeton University Press/ Project MUSE.

Goffman, E. (1959). *The presentation of self in everyday life*. Doubleday.

Guston, D. H. (2012). The pumpkin or the Tiger? Michael Polanyi, Frederick Soddy, and anticipating emerging technologies. *Minerva, 50*(3), 363–379. https://doi.org/10.1007/s11024-012-9204-8

Habermas, J. (1968). *Technik und Wissenshaft als 'ideologie'*. Suhrkamp verlag.

Habermas, J. (1971) *Knowledge and human interests*. Beacon Press.

Hacking, I. (1983). *Representing and intervening: Introductory topics in the philosophy of natural science*. Cambridge University Press.

Hacking, I. (1988). The participant Irrealist at large in the laboratory. *The British Journal for the Philosophy of Science, 39*(3), 277–294. https://doi.org/10.1093/bjps/39.3.277

Hacking, I. (1999). *The social construction of what?* Harvard University Press.

Hanson, N. R. (1958). *Patterns of discovery an inquiry into the conceptual foundations of science*. Cambridge University Press.

Hesse, M. B. (1972). Defence of objectivity. In *Proceedings of the British Academy, 58*.

Kitcher, P. (1993). *The advancement of science: Science without legend, objectivity without illusions*. Oxford University Press.

Kitcher, P. (2012). *Preludes to pragmatism: Toward a reconstruction of philosophy*. Oxford University Press.

Kuhn, T. (1962). *The structure of scientific revolutions. Second editio, enlarged 1970*. University of Chicago Press.

Labinger, J. A., & Collins, H. M. (Eds.). (2001). *The one culture? A conversation about science*. University of Chicago Press.

Lakatos, I., & Musgrave, A. (1970). *Criticism and the growth of knowledge*. Cambridge University Press.

Latour, B. (1987). *Science in action: How to follow scientists and engineers through society*. Harvard University Press.

Latour, B. (1993). *We have never been modern*. Harvard University Press.

Latour, B., & Woolgar, S. (1979). *Laboratory life: The social construction of scientific facts*. Sage Publications.

Lewontin, R. C. (2000). *It ain't necessarily so: The dream of the human genome and other illusions*. Granta.

Lewontin, R. C., Kamin, L. J., & Rose, S. P. R. (1984). *Not in our genes: Biology, ideology, and human nature* (1st ed.). Pantheon Books.

Longino, H. E. (1990). *Science as social knowledge: Values and objectivity in scientific inquiry*. Princeton University Press.

Longino, H. E. (2002). *The fate of knowledge*. Princeton University Press.

Menand, L. (2001). *The metaphysical Club*. Flamingo.

Merton, R. K. (1938). *Science, technology and society in seventeenth century England*. Saint Catherine Press.

Merton, R. K. (1968). The Matthew effect in science. *The Reward and Communication Systems of Science are Considered, 159*(3810), 56–63. https://doi.org/10.1126/science.159.3810.56

Merton, R. K. (1973). *The sociology of science: Theoretical and empirical investigations*. University of Chicago Press.

Miedema, F. (2002). Science fictions: A scientific mystery, a massive cover-up, and the dark legacy of Robert Gallo. *Nature Medicine, 8*(7), 655–655. https://doi.org/10.1038/nm0702-655

Miedema, F. (2012). *Science 3.0. Real science real knowledge*. Amsterdam University Press.

Misak, C. J. (Ed.). (2007). *New pragmatists*. Clarendon Press/Oxford University Press.

Misak, C. (2013a). *The American pragmatists The Oxford history of philosophy* (First edition. ed., pp. 1 online resource (xvi, 286 pages)). Retrieved from ProQuest. Restricted to UCSD IP addresses. Limited to one user at a time. Try again later if refused https://ebookcentral.pro-quest.com/lib/ucsd/detail.action?docID=1132322

Misak, C. (2013b). Rorty, pragmatism, and analytic philosophy. *Humanities, 2*(3), 369–383. https://doi.org/10.3390/h2030369

Misak, C. (2020). *Frank Ramsey: A sheer excess of power*. Oxford University Press.

Mitroff, I. I. (1974). Norms and counter-norms in a select Group of the Apollo Moon Scientists: A case study of the ambivalence of scientists. *American Sociological Review, 39*(4), 579–595. https://doi.org/10.2307/2094423

Nagel, E. (1940). Charles S. Peirce, Pioneer of modern empiricism. *Philosophy of Science, 7*(1), 69–80. https://doi.org/10.1086/286606

Nagel, E. (1961). *The structure of science; problems in the logic of scientific explanation*. Harcourt.

Perutz, M. (1995). The pioneer defended. *The New York Review of Books*.

Pinch, T. (2001). Does science studies undermine science? Wittgenstein, Turing, and Polanyi as precursors for science studies and the science wars. In H. C. J. A. Labinger (Ed.), *The one culture? A conversation about science*. The Chicago University Press.

Polanyi, M. (1958). *Personal knowledge: Towards a post-critical philosophy*. Routledge & Kegan Paul.

Polanyi, M. (1962). *Personal knowledge; towards a post-critical philosophy*. Harper Torch Books.

Polanyi, M. (1967). The growth of science in society. *Minerva, 5*(4), 533–545. https://doi.org/10.1007/BF01096782

Polanyi, M., & Grene, M. (1969). *Knowing and being: Essays*. University of Chicago Press.

Popper, K. R. (1959). *The logic of scientific discovery*. Hutchinson.

Popper, K. R. (1972). *Conjectures and refutations: The growth, of scientific knowledge* (4th ed.). Routledge & K. Paul.

Popper, K. R. (1976). *Unended quest: An intellectual autobiography* (Rev. ed.). Fontana.

Popper, K. R. (1981). The rationality of scientific revolutions. In I. Hacking (Ed.), *Scientific Revolutions*. Oxford University Press.

Putnam, H. (1981). *Reason, truth, and history*. Cambridge University Press.

Putnam, H. (1995). *Pragmatism: an open question*. Blackwell.

Putnam, H. (2002). *The collapse of the fact/value dichotomy and other essays*. Harvard University Press.

Putnam, H., & Conant, J. (1990). *Realism with a human face*. Harvard University Press.

Putnam, H., & Conant, J. (1994). *Words and life*. Harvard University Press.

Ravetz, J. R. (1971). *Scientific knowledge and its social problems*. Clarendon Press.

Ravetz, J. R. (2011). Postnormal science and the maturing of the structural contradictions of modern European science. *Futures, 43*(2), 142–148. https://doi.org/10.1016/j.futures.2010.10.002

Rorty, R. (1979). *Philosophy and the mirror of nature*. Princeton University Press.

Shapin, S. (1982). History of science and its sociological reconstructions. *History of Science, 20*(3), 157–211. https://doi.org/10.1177/007327538202000301

Shapin, S. (1988). Understanding the Merton thesis. *Isis, 79*(4), 594–605. https://doi.org/10.1086/354847

Shapin, S. (1994). *A social history of truth: Civility and science in seventeenth-century England*. University of Chicago Press.

Shapin, S. (1996). *The scientific revolution*. University of Chicago Press.

Shapin, S. (2002). Dear prudence. *London Review of Books, 24(2)*, dated 14 January.

Shapin, S. (2007). Science in the modern world. In O. A. E. Hackett, M. Lynch, & J. Wajcman (Eds.), *The handbook of science and technology studies* (3rd ed., pp. 433–448). MIT Press.

Shapin, S. (2008). *The scientific life: A moral history of a late modern vocation*. University of Chicago Press.

Shapin, S., Schaffer, S., & Hobbes, T. (1985). *Leviathan and the air-pump: Hobbes, Boyle, and the experimental life: Including a translation of Thomas Hobbes, Dialogus physicus de natura aeris by Simon Schaffer*. Princeton University Press.

Toulmin, S. E. (1953). *The philosophy of science. An introduction*. Hutchinson University Library.

Toulmin, S. (1972). *Human Understanding*. Clarendon Press.

Toulmin, S. (1977). From form to function: Philosophy and history of science in the 1950s and now. *Daedalus, 106*(3), 143–162.

Toulmin, S. (2001). *Return to reason*. Harvard University Press.

Watson, J. D. (1968). *The double helix: A personal account of the discovery of the structure of DNA*. Weidenfeld and Nicolson.

Ziman, J. (1968). *Public knowledge: An essay concerning the social dimension of science*. Cambridge University Press.

Ziman, J. M. (1978). *Reliable knowledge: An exploration of the grounds for belief in science*. Cambridge University Press.

Ziman, J. M. (2000). *Real science: What it is, and what it means*. Cambridge University Press.

Zuckerman, H. (1989). The other Merton thesis. *Science in Context, 3*, 239–267. https://doi.org/10.1017/S026988970000079X

Chapter 3
Science in Transition How Science Goes Wrong and What to Do About It

Abstract Science in Transition, which started in 2013, is a small-scale Dutch initiative that presented a systems approach, comprised of analyses and suggested actions, based on experience in academia. It was built on writings by early science watchers and most recent theoretical developments in philosophy, history and sociology of science and STS on the practice and politics of science. This chapter will include my personal experiences as one of the four Dutch founders of Science in Transition. I will discuss the message and the various forms of reception over the past 6 years by the different actors in the field, including administrators in university, academic societies and Ministries of Higher Education, Economic Affairs and Public Health but also from leadership in the private sector. I will report on my personal experience of how these myths and ideologies play out in the daily practice of 40 years of biomedical research in policy and decision making in lab meetings, at departments, at grant review committees of funders and in the Board rooms and the rooms of Deans, Vice Chancellors and Rectors.

It has in the previous chapters become clear that the ideology and ideals that we are brought up with are not valid, are not practiced despite that even in 2020 they are still somehow 'believed' by most scientists and even by many science watchers, journalists and used in political correct rhetoric and policy making by science's leadership. In that way these ideologies and beliefs mostly implicitly but sometimes even explicitly determine debates regarding the internal policy of science and science policy in the public arena. These include all time classic themes like the uniqueness of science compared to any other societal activity; ethical superiority of science and scientists based on Mertonian norms; the vocational disinterested search for truth, autonomy; values and moral (political) neutrality, dominance of internal epistemic values and unpredictability regards impact. These ideas have influenced debates about the ideal and hegemony of natural science, the hierarchy of basic over applied science; theoretical over technological research and at a higher level in academic institutions and at the funders the widely held supremacy of STEM over SSH. This has directly determined the attitudes of scientists in the interaction with peers within the field, but also shaped the politics of science within science but also

© The Author(s) 2022
F. Miedema, *Open Science: the Very Idea*,
https://doi.org/10.1007/978-94-024-2115-6_3

with policy makers and stakeholders from the public and private sector and with interactions with popular media.

Science it was concluded was suboptimal because of growing problems with the quality and reproducibility of its published products due to failing quality control at several levels. Because of too little interactions with society during the phases of agenda setting and the actual process of knowledge production, its societal impact was limited which also relates to the lack of inclusiveness, multidisciplinarity and diversity in academia. Production of robust and significant results aiming at real world problems are mainly secondary to academic output relevant for an internally driven incentive and reward system steering for academic career advancement at the individual level. Similarly, at the higher organizational and national level this reward system is skewed to types of output and impact focused on positions on international ranking lists. This incentive and reward system, with flawed use of metrics, drives a hyper-competitive social system in academia which results in a widely felt lack of alignment and little shared value in the academic community. Empirical data, most of it from within science and academia, showing these problems in different academic disciplines, countries and continents are published on virtually a weekly basis since 2014. These critiques focus on the practices of scholarly publishing including Open Access and open data, the adverse effects of the incentive and reward system, in particular its flawed use of metrics. Images, ideologies and politics of science were exposed that insulate academia and science from society and its stakeholders, which distort the research agenda and subsequentially its societal and economic impact.

3.1   The Royal Response (1)

In the fall of 2012, there were a few high-profile academic public events that were related to the discovery in the year before of a few serious fraud cases in The Netherlands in biomedicine and social psychology. The latter case was shocking and notorious for how it had been done with unflinching arrogance over many years. Because of its size and impact, it became worldwide known. I was present at the meeting held in September at the Royal Academy of Arts and Sciences where Kees Schuyt, a prominent sociologist and law scholar, as chair of a committee of the Royal Academy presented the advice that was focussed on responsible handling of research data (KNAW, 2012). The conclusions of the advice and of the meeting at the Royal Academy was that fraud and violation of the principles of integrity in research was believed to be very rare, but that it should be investigated. The feeling was that education of researchers about integrity, but also in the institutions technical proper handling of data should be promoted and enabled. Very cautiously the idea was mentioned of the obligation for researchers to making data available that supported claims in a journal paper to improve peer review. Finally, it was concluded that informal peer pressure in the community and in the later stages more formally through peer review should be improved. Despite a classical reference to the 'leading values of science which are distinct from any other social activity' and cautious conclusions, the committee did pose a series of critical questions that they believed should not be evaded. They suggested that the social system in which individual researchers do their work might allow or even invite misconduct. In that context they mention the incentive and reward system with its academic hierarchies and publication pressure (p60). The panel with members of Academy and Young Academy largely agreed. Of interest was the mentioning of some examples of serious fraud in physics (amongst others 'the Schön' case). In response to this, a very senior Royal Society member from the natural sciences remarked that of course this issue of quality is typical for *'the soft sciences and biomedicine, but not for us in the hard sciences, because in physics, through our experimentation, we ask a question to nature and nature gives a clear answer, so physics is beyond fraud'*. The chairman, a theology scholar who early in his career had become a professional university administrator, who knew about the problem of foundations, decided to let that one go. At the conclusion of the debate, I made a short critical remark from the floor, that something is really wrong with science if we focus on the rare fraud cases but are looking away from to the growing evidence of a large 'grey zone' of shoddy science, also in other disciplines than biomedicine and social psychology. This grey zone is not populated with fraudsters or bad people who are to blame, but honest researchers that try to survive in our crazy academic system driven by perverse incentives and rewards. This I thought should be acknowledged and discussed. What I had in mind then was in fact to become one of the corner stones of Science in Transition and of this book. The chairman's reply was 'that may be so, but we cannot change a whole system' and then there were drinks, gossip and appetizers (typically Dutch 'bitterballen') in the foyer.

Kees Schuyt was interviewed in a national newspaper and to my relief was much more open about the likely systemic cause of the problems. In October at a meeting held in Spui 25, a University of Amsterdam open podium/debate centre, Huub Dijstelbloem took part in the panel discussion with Kees Schuyt and Andre Knottnerus an authority in the Dutch health science and governmental science advice system. The debate was much more open and critical and did not evade the problems of the system.

November 28, 2012, at the Royal Academy again, a committee chaired by Pim Levelt, a former President of the Academy, presented its investigation of fraud and misconduct of Diederik Stapel. (https://www.rug.nl/about-us/news-and-events/news/news2012/stapel-eindrapport-ned.pdf). This case, together with a case at Erasmus Medical Centre, since their discovery in September and December 2011, dominated the debate about trust in science in the country. The committee revealed the technical and methodological aspects of the case in great detail. In their final comments they state that *'Committees that have evaluated the research of social psychology, have not recognized some of the signals that the committee in this report do describe. They simply were relying on peer review both with respect to methodology and contribution to theory. Another issue in this context is to what degree these evaluation committees are instrumental in sustaining the assumed undue publication pressure and connected mores and behaviours. This specifically concerns requirements of numbers of publications, the order of authors, responsibilities of co-authors and repeated publication of similar results.' (translation FM).*

The Science in Transition Team

A year later in November 2013, the public start of *Science in Transition* took place at the same prestigious venue of the Royal Academy of Arts and Sciences on one of the channels in the centre of Amsterdam. The *Science in Transition* team started its work in January 2013. Huub Dijstelbloem, whom I already mentioned, had the years before been very active in national debates about incentive and rewards focussed on inclusive indicators and methods for evaluation of the impact of research. He also studied public participation and policy making which is discussed in Chap. 5. The other three members of the group that started *Science in Transition*, were Jerome Ravetz and professors Frank Huisman and Wijnand Mijnhardt. The five of us did not really know each other, but we shared our thinking about science which brought us together.

Jerome Ravetz (1929), Jerry, as we call him, replied promptly and enthusiastically, full of energy looking for action when I had send him in the fall of 2012 my little book about science, *Science 3.0, Real Science, Significant Knowledge* (Miedema, 2012). I did not know him, but knew his 1971 book (see Chap. 2). Ravetz with a small group of colleagues had published in 1993 a paper in which they described another way of doing science, explicitly with the aim to deal with policy issues of high risk and high uncertainty for science is critical but for which the time for deliberation is limited. They coined the name *Post-Normal Science* for an approach in an integrated and democratized process in which all relevant knowledge and social values and the relevant publics are fully acknowledged and participate (Funtowicz & Ravetz, 1993). In the months that followed Jerry received a Fellowship of the Descartes Centre of Utrecht University which brought him and his wife frequently to Utrecht. His first visit was to Amsterdam on January 4, 2013 when we talked the whole day and part of the evening about his work, his thoughts about science in 2013 and the actions to be taken.

(continued)

Frank Huisman, Huub Dijstelbloem and Jerome Ravetz. (Amsterdam, February 2013)

Frank Huisman (1956) is at Maastricht University, the interdisciplinary group of Science, Technology and Society Studies (MUSTS) and since 2006 full professor of the History of Medicine at UMC Utrecht. His interest is the history (and sociology) of modern medicine. Together we started in 2009 a selective advanced PhD course on philosophy and sociology of science, called *This thing called Science*. The course proved to be an immediate success: 120 PhDs applied to be enrolled in the course which offered place to only 45 of them, and students declared it to be the best course offered by the Graduate School of Life Sciences. There was clearly a great need among some PhD students to learn about the history, philosophy, ethics and politics of science, and be socialized into the biomedical sciences in a different way. We felt very happy to be able to create this new awareness among a new generation of biomedical researchers.

Frank Huisman introduced Wijnand Mijnhardt (1950) to me at the end of November.

(continued)

Wijnand Mijnhardt is an international well-known historian of culture and science and at that time served as Chair of Comparative History of the Sciences and the Humanities. He is founder and past director of the Descartes Centre for the History and Philosophy of the Sciences at Utrecht University. I told Wijnand I was honoured that he came to my room and I pulled his leg stating that '*his Centre, to the best of my knowledge, preferably studied scientists and scholars that had already passed away a long time ago. I understood, I said, that this nicely avoids the political issues that in our time trouble academia and society, but*', I said, '*my goals are quite the opposite. Our thinking about science should, in the good tradition of pragmatism lead to action in the real world to improve the academic lives of our stakeholders: graduates, post-docs, students and professors in our universities and those in society alike*'. Wijnand appreciated the humour and loved this idea for the project. He in the following years eloquently brought to the table his strong opinions with colourful flavours in the context of *Science in Transition*.

Sarah de Rijcke, photo taken by Bart van Overbeeke

Regarding the composition of the team, we were criticized and had to admit, that we had a problem: we were five, and later four, older white males who each had done well in the system. This was corrected in part very soon, when Sarah de Rijcke a well-known researcher in STS at CWTS Leiden and an expert on all the issues Science in Transition was addressing joined the team. We did not have graduate students or early career scientist in the team. Our best defence to this was that, given the idea that changing a social system goes against the elites and the most powerful in that very system, we were not vulnerable to the classical framing of 'being a couple of losers complaining about the system in which they had failed'. Many who question the mores and rules of the system indeed are told 'If you cannot stand the heat, get out of the kitchen'.

Science in Transition

We started from the optimistic, some thought naïve, perspective of the possibility to improve science. Our analysis of the problem had a broad scope, from quality issues, fraud, poorly conducted or irrelevant science, agenda setting and responsiveness to issues in society, assumptions, ideologies and hierarchies that distorted the system internally in academia and in interaction with society. We wanted at all cost to avoid the well-known type of general academic discussions about problems of 'the university' and easy blaming of 'incompetent' administrators, lazy students or neoliberal economics'. Angry complaining and blaming without realistic directions for improvement would stifle our initiative like has happened to many initiatives before. For all of us it was clear that these problems had to be approached in a larger context of the socioeconomics of the institutional organization of science. From the start it was clear that we needed to more specifically discuss the contribution of the incentive and reward system. Persistence of specific problems, in our view, seemed to be related to the system of research evaluation in institutions and by funders as it had gradually developed since the 1980s. Our focus was very much on research, but in the incentive and rewards structures in academia this related to the poor appreciation of teaching and teaching careers and so this was discussed as well.

All of these issues one by one were not new, but we believed that an integral approach of the issues seen as parts of one social system was going to be quite unique. Many science writers and philosophers discussed their favourite views and worries, but a consistent system approach to science to our knowledge was very rare, if available at all. Apparently, without such an explicit awareness, we felt confident as a team to have enough complementary experience in science and academia, both in theory and practice, to take on this ambitious project. We decided that we had to get a proper analysis and comprehensive picture first. We agreed that going from there, to achieve long-lasting improvements, concerted actions of the community were required. This involved systemic institutional change in which academic leadership at universities, especially from Rectors, Deans, Royal Academies and prominent scholars, public and private funders should be engaged and committed.

Three workshops were held on *Image and Trust, Quality and Corruption, Communication and Democracy* in April, May and June 2013. Next to the initiators, about ten scientists from the Netherlands were invited to participate in each of the workshops. Participants were hand-picked by us, known for their expertise, critical thinking and outspoken views about science. (See website scienceintransition.nl for the lists of participants and the workshop presentations.) Based on the results of these three workshops in the summer of 2013, a draft position paper was produced by the initiators, mainly by exchanges via the mail.

3.2   The Royal Response (2)

We were not alone in this endeavour. In the Royal Society, in that same period, a committee was working on a report on trust in science. With reference to the recent high-profile fraud cases, the Ministry of Education, Culture and Science (OCW), in January 2012 had formally asked the Royal Academy to advise the Ministry regarding trust in science. In it there was a specific request to advice on possible actions by the main actors in the domain of science, including research institutes, funders and government, that could help improve integrity and trust. This committee started in March 2012 and published their advice in May 2013. The report was presented in May 2013 by the committee chaired by Keimpe Algra, a humanities scholar, who was to become Dean of the Faculty of the Humanities of Utrecht University the next year (KNAW, 2013). The committee in response to the questions from the Ministry had taken a broad approach, explicitly including the wider system and community of science. They concluded that the Mertonian rules were under pressure because of changes in the scientific system that had occurred in the past 30 years, with consequences for the practice of research. This agrees with the analysis of the legacy of Merton presented in Chap. 2. From this analysis the committee concluded that there is, with respect to integrity and quality, a duty for the individual to show '*honesty about the research goals and intentions*'. At the same time and with even more emphasis the institutions, universities and funders were urged to take responsibility for the culture of science where it did not promote or even obstruct proper behaviour and integrity of researchers. A couple of times the committee suggested that it would be a good idea to have institutional accreditations for research to help the institutions in setting up and uphold the relevant practical policies. The example of quality assurance policies in health care were mentioned. Although the committee was cautious with respect to top-down programming of research, they made it clear that not only should research be done right, but also the right research should be done, which brought 'agenda setting' and external values as a novel dimension in the discussion. It was proposed to invest in more practical awareness and social control in the form of positive peer pressure with important roles for the research communities in universities and research institutes. For this, they said an open and safe academic culture was required.

The committee, in contrast to previous reports discussed the problematic effects of the external forces on the practice of academia. This related to allocation of funds and collaborations with private commercial partners. The increasing influence on academic life also of tenured staff, of short-term funding schemes, being focussed on 'sexy topics and hypes' and the researcher's temptation to promise unrealistic impact and novelty were mentioned as a distortion of the dynamics of academia. This induced bias against replication and negative results, also at the journals, goes against long term more difficult research. The committee states that this is reflected in national research evaluations which enforces these practices and on a focus on numbers of publications. Concluding with constructive suggestions, the committee did not suggest to really rethink the issue of cause-and-effect regarding the problems discussed. They were nearly there but did not take that logical next step to conclude or at least suggest that the institutional organization with its incentive and reward system, and specifically its indicators for excellence, critical for decisions on funding and career advancement might provoke strategic behaviours that caused or at least promoted many of these interdependent problems.

This advice, *'Trust in Science'*, was discussed at a meeting at the Royal Academy in September 2013 where I was invited to give a talk and presented the Science in Transition Position Paper and *'A Toolbox for Science in Transition'* to reassure the audience, of mainly early career scientists, that national and international change was possible (Supplement 1).

3.3 Science in Transition Position Paper, October 2013

A final version of the Position Paper, incorporating comments that we had received thus far was published on the website on October 17, 2013.

https://scienceintransition.nl/en/about-science-in-transition/position-paper

The Position Paper is composed of the chapters: Images of Science, Trust, Quality, Reliability and Corruption, Communication, Democracy and Policy, University and Education, and a brief Conclusion paragraph.

3.4 Science in Transition: A Systems Approach

Science in Transition as an initiative and movement to improve the impact of science and research entered a field where many had gone before. We were heavily inspired and influenced by many different scientists and scholars who had been writing about science and society, as the reference list of the Position Paper duly reflects. These writings and actions go back to the 1970s and deal with the ideology of science and its Legend, the sociology and social organization of science and with problems of science in and with society. In the years before 2012, major initiatives with respect to quality and reproducibility in research had started. This in reaction to the increasing evidence from empirical research showing poor quality and unexpectedly low reproducibility in biomedicine and psychology, but also other fields of research (Altman, 1994; Begley & Ellis, 2012; Prinz et al., 2011) (Ioannidis, 2005; Ioannidis et al., 2012; Moffitt et al., 2011; Moore et al., 2017; Nosek et al., 2012). I will not discuss the history of this meta-science work on poor quality research and replication, as that has been done by experts before. Our interest in the context of *Science in Transition* was to understand why poor research is being done and published. It had not been decreasing even as we know about it but had apparently rapidly been increasing to be a problem in the more recent years. Many studies had already shown the relation between the use of bibliometric indicators in research evaluation, thus in the incentives and rewards system and strategic behaviour of researchers (Hammarfelt & de Rijcke, 2014; Moore et al., 2017; Wilsdon, 2016; Wouters, 1999, 2014; Wouters et al., 2015).

It is this problem that is, more implicitly though, addressed by the San Francisco **D**eclaration **O**n **R**esearch **A**ssessment, known by its acronym **DORA**, that started in December 2012:

> There is a pressing need to improve the ways in which the output of scientific research is evaluated by funding agencies, academic institutions, and other parties. To address this issue, a group of editors and publishers of scholarly journals met during the Annual Meeting of The American Society for Cell Biology (ASCB) in San Francisco, CA, on December 16, 2012. The group developed a set of recommendations, referred to as the San Francisco Declaration on Research Assessment.

The debate about quality and impact of research in biomedicine reached a novel international level in January 2014 with a series of articles under the heading *Research: Increasing Value, Reducing Waste* in the *Lancet* on January 8, 2014. This was the result of an initiative of a group of very established biomedical researchers, who in some case already for many years had focussed on quality issues related to methodology, design and problem choice in clinical studies in humans, but also in animal studies. Internationally the best known are: John Ioannidis (METRICS Stanford University), Doug Altman (Oxford University), Ian Chalmers (Oxford University) and Paul Glaziou (Bond University) (http://www.thelancet.com/series/research) and the initiative is called *Reward Alliance* (http://rewardalliance.net/increasing-value-reducing-waste/).

In the same month, a paper was published in *Nature* by the NIH leadership, Francis Collins and Lawrence Tabak, announcing the NIH reproducibility project. This project is an adequate reaction to the seminal study by Begley and Ellis published 2 years before also in *Nature* on poor reproducibility of pre-clinical biomedical research published in *Nature, Science* and *Cell* and an earlier study by Prinz et al. This was boosted in March 2012 by a paper in PNAS with the ominous tittle Rescuing *US biomedical research from its systemic flaws, w*ritten by five very high-profile authors, from the US biomedical science community. The best-known authors included Bruce Alberts, a former long-serving editor of *Science,* Shirley Tilghman a former president of Princeton and Harold Varmus, a former Director of NIH, former president of Memorial Sloan Kettering Cancer Centre and then director of the National Cancer Centre of NIH and last but not least the 1989 Nobel prize winner (https://www.pnas.org/content/111/16/5773).

2013
- San Francisco **D**eclaration **O**n **R**esearch **A**ssessment: stop using bibliometric indices for evaluation of researchers
- NIH: Reproducibility initiatives in life sciences and psychology
- Economist: How Science Goes Wrong, Trouble at the lab

2014
- Lancet: increase value and reduce waste in biomedical research
- Nature NIH, F. Collins & L.Tabak: increase reproducibility and change academic incentive system.
- PNAS March 2014 Alberts, Varmus et al. Rescuing US biomedical research from its systemic flaws
- Nobel prize winners **Schekman** and **Brenner** call for change, away from quantity and impactfactormania

---

3.5   How Scientists Get Credit

This discussion about quality of reporting and actions to be made to improve science was, in agreement with the initiators' professional backgrounds, for years mostly focussed on methodology, statistics and trial design. However, with these papers in different so-called prestigious 'high impact factor' journals, that attracted quite some international attention, the discourse broadened to take in to account another critical aspect. To the best of my knowledge, for the first time the distorting systemic effects of research evaluations were explicitly mentioned in public debates and discussions in academic circles. Indeed, that is the most dangerous of 'elephants in the room' of science and academia, that almost all writers of papers, policy reports and Royal Academy advisors about trust and quality had evaded.

We, in Science in Transition, were convinced that without including in our analyses and actions this crucial part of the system, little progress can be expected. It was a corner stone of the Position Paper and we have argued strongly for it, although the critique was that it would be impossible to change because many different players with divergent and contrasting interests are involved. Most of the problems that were pointed out by *Science in Transition* and the national and international initiatives described above, may at least be maintained or even institutionalized by the incentive and reward system. For some time now since 2013 the 'metrics' the use or in fact abuse of bibliometric indicators has been a central issue. We had the fortune to have Paul Wouters, an international distinguished researcher in the field of bibliometrics in our team and soon as remarked before, Sarah de Rijcke affiliated with CWTS joined the team. Paul Wouters was in 2012 appointed as Director of the Centre for Science and Technology Studies (CWTS) at Leiden University.

Use and Abuse of the Metrics

The five initiators of *Science in Transition* have been introduced, and I will introduce some of our fellow travellers with the progress of the narrative. From the summary of Paul Wouters' contribution to the second workshop it is clear that his expertise and broad experience with both theory and practice of bibliometrics and of the social organization of science were of utmost importance. In the public debate, Paul was very visible and strongly connected with the *Incentive and Rewards* theme of *Science in Transition.* Paul appeared to have a very interesting and colourful career. He has a Masters in biochemistry (Free University of Amsterdam, 1977) and a PhD in science and technology studies (University of Amsterdam, 1999). His PhD thesis titled *"The Citation Culture"* (1999) is on the history of the Science Citation Index, on and in scientometrics, and on the way the criteria of scientific quality and relevance have been changed by the use of performance indicators. In between these degrees he has worked as science journalist and as editor-in-chief of a daily newspaper ("De Waarheid"). This newspaper was the daily newspaper of the Dutch Communist Party (CPN) that in 1990 was stopped when the CPN merged with the Green Left political party. From 2010 to 2019 he was Director of The Centre for Science and Technology Studies (CWTS). From 2016 on, Wouters served on several EU expert groups that were started by the DG Research and Innovation to advice on the transition to Open Science. Since January 2019, Paul Wouters is Dean of the Faculty of Social Sciences, Leiden University. (Source and citations: from The Leiden University website)

https://www.universiteitleiden.nl/en/staffmembers/paul-wouters#tab-2

In the second workshop of *Science in Transition* held in June 2013, Paul Wouters gave a seminar largely based on his article, *The citation from culture to infrastructure,* which was published later that year (Wouters, 2014). Wouters presented an overview of his own work and major studies of other bibliometricians on the different types of effects of research evaluation, and specifically the use and abuse of bibliometric indicators. As virtually every debate about Incentives and Rewards is still dominated by the use and abuse of metrics this appeared to be a corner stone of the analyses that we did in the context of *Science in Transition.*

In the 1960s, at the advent of bibliometrics, its focus was on studying the dynamics of the different fields of scientific research to help understand nearly real time where science and the scientist are going. To know what the scientist are working on, what the big questions are in the different fields and, also of interest, what they do not (yet) study. Dynamics was in terms of changing numbers of papers and authors and thus researchers and funding. Tracking citations and citation patterns was done to discover networks of researchers working on somehow related problems and the relative importance of specific research questions based on citations to that work. Paul Wouters has studied the history of the *Science Citation Index (SCI)* which was developed and launched by Eugene Garfield in the early 1960s. I remember them from my first visits to NIH, and readers of my age will remember these enormous yellow SCI books in the library. They allowed you to track who had recently cited your papers and which papers of colleagues and competitors were cited or not. It took some time for the SCI to be used by a larger fraction of the community other than bibliometricians. *'This use increased markedly'*, as Wouters wrote in 2017 his obituary of Garfield, *'after the Journal Impact Factor was marketed in the SCI Journal Citation Reports starting in 1975'*. Garfield like many other bibliometricians with regard to JIFs *'was uncomfortable with their misuse as performance indicators'* (Paul Wouters, 2017).

The bibliometricians, were not naïve, they did not believe that science was guided by a neutral 'invisible hand' or Polanyi's autonomous 'Republic of Science'. The use of 'their' indicators in the evaluation of research, research institutions, and even at the level of individual scientists as performance indicators, was unwanted and mostly incorrect use of their work. From the early papers on this issue one sometimes gets the impression that they had not in full anticipated this cross-over between bibliometrics and sociology of science and later research management and research governance. In that cross-over, the indicators were used not to understand the dynamics of research by looking back at its recent past, but to steer and manage the direction and the agenda of research in a forward-looking approach (Whitley & Gläser, 2007). This had wide ranging effects outside academia. Wouters argued that there is convincing evidence that worldwide, since the 1980s research evaluations, and the indicators employed, to a great extend shape the research agenda ('problem choice') at universities and funders. The greatest impact of these performance assessments is, as will be discussed below, the direct effects it has on the allocation of research funds at the EU, national and university levels. This was until 2000 relatively rare, but he cites the paper by Diane Hicks on this saying that: *'By late 2010, 14 countries had adopted a system in which research funding is explicitly determined by research performance* (Hicks, 2012). Examples of indirect effects of performance indicators have been described, for instance the Standard Evaluation Protocol (SEP) in the Netherlands. In the SEP evaluations with comparisons of research done in similar fields of research in all universities have been done which has since 1990 increasingly been based on quantitative metrics. In this national research evaluation, funding is not directly distributed based on such rankings but the effects on reputation, esteem and standing in the field are well recognized and anticipated (Van der Meulen, 1997). It thus has become common practice to show in resumes a publication list with JIF's and the most current h-index. The latter has since its launch in 2005 (Hirsch, 2005) seen a very rapid and world-wide use which as Wouters said *'makes the h-index itself an indicator of indicator proliferation'*.

Even more important, in a natural human reflex, researchers anticipate the use of these metrics when their work will be judged and started to show several strategic behaviours. It has been shown that when evaluations focus on numbers of articles and on the JIF, that is the venue where these articles are published, this will be directly affecting the type of output of the researchers. If books, or articles in professional journals or publications in the national language are not valued, and don't score in the system, despite intrinsic interest or possible impact they will have for specific fields, researchers will undertake much less efforts in that direction (Butler, 2007; Laudel & Glaser, 2006). Even more important is the effect on the choice of research topics and engaging in multidisciplinary work. As will be argued and demonstrated in later chapters, the use of metrics can have detrimental effects on types of more applied research which has large and urgent societal impact but does not bring credit points to the researchers because results do not get published in journals with a high JIF. Active researchers, dependent on grant money, recognize these survival mechanisms immediately and most of them as we shall discuss later obviously find this very frustrating. When talking to university administrators, board members of academic medical centres and directors of funding agency's many times I heard them say in all honesty that this description of the behaviour of researchers must be a gross exaggeration. One may safely assume, they said, that the behaviour and choices of our highly educated scientists and staff are not likely to be so simply influenced by these metrics and indicators. Having been on committee's with scientists that evaluate resumes for academic promotions or grant proposals this behaviour is explicitly visible and audible, both by the committee members and by the candidates and from the materials under review.

In the final paragraph of the 2014 paper, Wouters says that it has to be seen how in the future these behaviours will develop and whether they will persist. As we know now, they did persist and are used to rank the universities world-wide (Hazelkorn, 2011) The topic of the perverse effects of the abuse of metrics and how they invite or even enforce strategic behaviour of scientists was and is still hot. This is sad but not unexpected given the analyses in this book that show how hard it is to change the indicators in order to make the system more inclusive, qualitative and fair.

Wouters as Director of CWTS was prominently interviewed by NRC, Trouw and Volkskrant in the months after the first symposium. In May 2014, together with John Ioannidis, we were interviewed in an article about incentives and rewards in "Medisch Contact', a Dutch weekly widely read by the medical profession published by **The Royal Dutch Medical Association (RDMA)**.

At CWTS, Paul Wouters and colleagues had launched a research programme in 2012 that took the role of metrics in science head-on. Paul Wouters and Sarah de Rijcke were also members of the team that wrote during 2014 and 2015 a thorough *'Independent Review of the Role of Metrics in Research Assessment and Management'* in the UK Research Excellence Framework, with the title *'The Metric Tide'* (Wilsdon, 2016). This report, came on the basis of broad and detailed research, to quite similar conclusions as *Science in Transition* but had a strong focus in addition on scholar publishing and bibliometrics. In the accompanying literature review, new empirical studies are cited that paint a detailed picture of how metrics are not only being taken up in research management and decision-making, but also feed into quite run-of-the-mill choices scientists make on the shop-floor: metrics-infused decisions that structurally influence the terms, conditions and content of their research (Rijcke et al., 2015; Wouters et al., 2015). Around the same time, Paul Wouters and Sarah de Rijcke, then both at CWTS Leiden, with three colleagues among which Diana Hicks, published *The Leiden Manifesto* (Hicks et al., 2015). It calls for a different way to evaluate research based on an inclusive set of ten principles (Supplement 3).

Entering the Field (1)

I did bench research on bacterial vaccines as my military conscription, at The Netherlands Institutes of Health (RIVM), just outside Utrecht. In that way as a researcher, I entered the field of science and was pretty much immediately introduced to the credit cycle. Being trained in immunology in Groningen, at RIVM I joined a small research team that made a new type of experimental (so called conjugated) vaccine ragainst bacteria (*Neisseria meningitidis*) that cause disease among children and young adults. The compounds were tested for induction of a protective immune response in mice. We did in that year 1980, quickly a lot of experiments with nice positive results that were presented at meetings and published the years after I had already left, with me being a second or third author. The senior investigator had set up a very productive and original collaboration with a high-profile pyrolysis mass-spectrometry group at AMOLF in Amsterdam, then 'the heaven of physics' in the Netherlands. A grant was obtained for me to pursue our work as a PhD student. AMOLF wanted to do more biophysical life science research, so I did a job interview in the fall of 1980 with the director. I was quite nervous since the director was professor Jaap Kistemaker, an impressive man, well known for having developed the principle and technology of uranium enrichment with ultracentrifugation. I was offered the job but chose not to join AMOLF which I had to tell Kistemaker over the phone. I preferred a job offer per January 1981, in a PhD position at CLB in Amsterdam. CLB then was regarded as one of the finest immunology institutes in the country. Kees Melief, a MD PhD then in his early forties, was heading the unit. He had returned from the US after a very productive stay in Boston. He had published well and was considered one of the new generation biomedical scientists with strong vision and a modern view of science and research. Melief took with him the American research culture and knew how to lead his team to the top of the field. We were a modern immunology lab where consciously strategic choices were being made on what to study. We were closely following the international fronts of the field and developments at the funders. In those days funding for biomedicine was rapidly increasing on the promise of new insights from molecular biology. The lab culture was to drive for results and publications and was already aimed at the top journals and we played the journal impact factor game. We were conscious of national competition and competitors abroad and thus highly competitive. New technologies like the generation of monoclonal antibodies, molecular biology, oncogenes, novel methods in molecular virology and immunology were immediately incorporated, experiments were done in the mouse and with blood samples obtained from patients.

(continued)

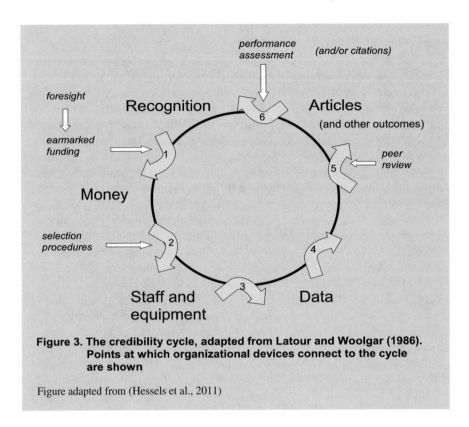

foresight

performance
assessment   (and/or citations)

earmarked
funding

selection
procedures

peer
review

Recognition

Articles

(and other outcomes)

Money

Staff and
equipment

Data

**Figure 3. The credibility cycle, adapted from Latour and Woolgar (1986).
Points at which organizational devices connect to the cycle
are shown**

Figure adapted from (Hessels et al., 2011)

3.6   The Credibility Cycle: Opening Pandora's Box!

Despite being heavily criticized by papers like *The Metric Tide* and, *The Leiden Manifesto* and despite high profile and widely endorsed actions from within the community of science, such as DORA, the abuse of metrics clearly still is common practice around the globe. To understand this persistence of the use of metrics, we have to understand the role incentives and rewards, and of the critical role of metrics therein in the institutional and social organization of science and academia. As Paul Wouters argued, since the 1950s the sociology and management literature on the institutionalization and organization of science - in academia, universities and the various funding organisations- mentions incentives and rewards as an important component of the governance of the community of science. As discussed in Chap. 2, that literature was too respectful of science and even more of scientists and was normative. In the early days from Merton to Popper and Polanyi, science was believed to be at the discretion of the community of science, interference was not appreciated. The reward system was part of the *'Black Box of Science'*, not to be questioned by outsiders, who anyway were believed to not understand science at all.

This Black Box was part of the magic of the 'science knows best' narrative of Vannevar Bush after the Second World War. Especially basic science, which also goes by the names of 'blue skies', 'curiosity-driven' or 'free science', when left alone, of course well-supported, so it goes, will have huge returns for society, for the military and economy. As discussed in Chap. 1, this was the power of the marketing and sales of natural basic science between 1945 and 1960 and the basis for the distinction between the natural and biomedical sciences and the humanities and the social sciences.

In the late 1950s, Peter Winch, in his *The Idea of a Social Science*, was the first to point out that the research in social sciences and the humanities is also science, albeit it a different form of the natural science and research and should not be judged by the frame of 'the scientific method' of the natural sciences. (Winch, 1958) Winch as many others, apparently was still under the impression at least left it open, that the natural sciences indeed were successful because they had a unique formal, well-founded and infallible method. This is, as I discussed in Chap. 2, in the positivist context of those days, not strange. The 'successes of the natural sciences' were the main reason why it was in these days, but still, mind-boggling even for most philosophers to admit that even in the natural and biomedical sciences there is no general, validated, formal and universal timeless method. I already pointed out that Ernst Nagel in his influential textbook of 1961, discussed this problem in general terms as well as other methods of inquiry appropriate for the social sciences (Nagel, 1961).

As we have seen when philosophers, historians and sociologist after 1970 started to study science as practice, eventually they also came closer to the social system and the Black Box that hid the rewards system from the eyes of outsiders. Stephen Toulmin (1972), John Ziman (1978), and a few other authors, after Winch, explicitly expressed a conceptual critique regarding the reward system of science and the indicators used in research evaluations in comparisons between academic disciplines. This was at that time not yet about the type of metrics, but as discussed in Chap. 2, about the myth of the method of the natural sciences in contrast to the hermeneutics (interpretative methods) and reasoning ('the vague methods') of the humanities and the social sciences. As a consequence of this belief of the supremacy of the method of the natural sciences, the social sciences and humanities, these authors concluded, were systematically undervalued. They were getting a bad deal in academia. Toulmin was one of the first in his *Human Understanding* (Toulmin, 1972) to take this insight to the 'corridors of power' of academia and firmly attacked the positivist Cartesian dominance of the natural sciences and to point to it as the cause of this unequal fight between these disciplines in academia. He believed this was the poverty of academia, a major problem for the enterprise of science and scholarship in society. Ziman in his early work criticizes the ideology of the legend and the natural sciences but struggles with the idea that SSH have their own field of inquiry with proven methods and huge impact. It is of interest to note, as I did in Chap. 1, that only 13 years before C.P. Snow had criticized the humanities for their snobbery regarding the natural sciences (Snow, 1993).

We see the connect between the Legend, its philosophy of science, with the way science became organized and governed since 1945. From the 1960s on, but definitely in the past 40 years a multitude of complex often antagonistic interactions between society, academia, universities and knowledge institutes has shaped science in all possible meanings of the word 'science'. In these interactions, communication, debates and conflicts, contracts and agreements, serious power relations are at play that shape science and the growth of knowledge at many levels. This involves science as the national and global system of public knowledge production, science as the total of disciplines organized in the structures of academia, including the sciences and social sciences and humanities (Guston, 2000; Rip, 1994; Rip & van der Meulen, 1996; Whitley, 2000).

3.7 Distinction

At the institutional level, virtually the whole of academia became organized by a social system that is most adequately described by Bourdieu's concept of 'a field' (Bourdieu, 1975, 2004). It is a truly social game of stratification, elites and distinction based on indicators about professional quality and excellence but also on habitus and subtle social rules. We have seen in Chap. 1, that the idea of 'pure' and 'applied' science has been and still is an ideological concept that is, by both sides, called upon in debates about science policy. Bourdieu, in his seminal book *"Distinction'* published in 1979 in French and in English in 1984, provides amazing insight and understanding of the different cultural, political and social tastes and preferences of the two main social classes (Bourdieu, 2010). Based on empirical sociological research performed in France in the 1960s, this is primarily discussed for tastes of the arts, painting, literature, furniture and music. The ideas of 'pure', 'abstract', 'universal', 'disinterested', 'distance to necessity' are indicators of the distinction of 'high culture'. It is clear that 'this distance to necessity' that provides economic freedom for useless and free thinking is a privilege of the middle and upper class. Bourdieu shows how members born in families of these economic, socially and culturally distinct classes fare in education and academia. Building on these insights, and the concepts of habitus and field, a host of research has shown that this is not typically French. In Chap. 1 I already discussed the influence of class distinction in England on the preference for pure over applied science which was 60 years go criticised by C.P. Snow and Peter Medawar (Medawar, 1982; Snow, 1993).

This schism historically and philosophically runs deep. First Plato, of course, is mentioned by Bourdieu as a source (p47), but in a postscript (p487–502) this distinction between 'pure' and vulgar' is taken to philosophy and academia with many citations from Kant's *Critique of Judgement*. Indicators of high culture relate in the Greek classical natural philosophy to the opposite of *charming, easy (pleasure and listening), facile, bodily pleasure, common (as in common knowledge).* 'Pure' thus suggest more difficult, requiring more perseverance compared to 'applied' which is crude, easy and with results to be readily obtained. The *'taste of reflection'* is opposed to *'the taste of the senses'.* I like to use 'high church' versus 'low church' to designate this distinction.

Five years after *Distinction,* Bourdieu published *Homo Academicus* where he studied how citizens born in different social classes achieve in the respectively preferred educational trajectories leading to academia with distinct preferences for specific faculties and jobs in and outside academia (Bourdieu, 1988). Finally, Stokes presented in his *Pasteur's Quadrant,* a critical survey of the idea of pure and applied science in relation to technological innovation (Stokes, 1997). Stokes discussed how since the times of classical Greek philosophy, philosophy per definition ought to be 'pure' and not deal with mundane and real-world problems. He cites A.C. Crombie saying *'it remained characteristic of Greek scientific thought to be interested primarily in knowledge and understanding and only very secondarily with practical usefulness'.* (p29) Stokes shows that this idea of pure science and research, next to the rise of technology and applied science since Bacon in the nineteenth century, survived with sharp ideological and organizational separations in the system in mainly France and Germany.

The 'pure' and 'applied' distinction, like the schism between the 'hard' and the 'soft' sciences, has been to a great extent adopted world-wide and is very much alive within the natural sciences and biomedical research, but also within the social sciences and the humanities and has in the past 40 years in addition been institutionalized by the corresponding metrics. It still comes with the whole connotation of professional scientific but also political and cultural distinctions of 'high' and 'low church' and, as described by Bourdieu, cannot be underestimated as part of the power games of the academic field. If a scientist explains that the does fundamental or basic science, this implicitly but really means to say that he or she in his or her field belongs to the class of scientist with highest reputation and highest standing. During the Covid-19 crisis experienced scientists from all academic different disciplines spontaneously started research in multidisciplinary teams to fight the virus and its public health, social and economic crises. Virtually all of the scientists that we saw in the media and who did the work, most of their professional lives did research on for instance biology, epidemiology or mathematical modelling in the applied context of infectious diseases. Still, scientists from the 'hard' and 'pure' sciences argued that COVID-19 had demonstrated again that it was **fundamental** science that made major contributions in our dealing with the crisis, basic science should receive increased funding. Mind you, in most cases basic science in this type of political statements refers to basic natural sciences.

In the power struggle to enter a field and for upwards mobility, indicators and criteria for excellence are employed within science not by voting or a democratic process, but by colleagues (peers) in committees, advisory boards and promotion

Advancing in the Field (2)

We learned 'science the modern way' by doing. We learned how the write, how to present and how to do our networking at meetings. We learned by looking on how Melief organized the lab, how he was critical regards novelty, rigor and quality, played the game of networking and publishing, how he dealt with peer review, and wrote his grants. In the days before the internet, we combined meetings in the US with visits to relevant labs to present our work. My first roundtrip was in December 1982 with visits to Mount Sinai NY, NIH/NCI at Bethesda, Stanford and a cellular immunology meeting at Asilomar, near Monterey and a laboratory at UCSF. Melief showed us how to move, discuss, pointed out the competition, criticized bad talks and introduced us to famous colleagues.

Melief Lab retreat Spring 1981

At Asilomar, December 1982

Grants, we learned, have to deal with the short cycle and be risk avoiding. You should pick problems that are considered relevant but should not be too complex or too difficult. If the grant is received, after the typical 4 years the

(continued)

grant is running you must have something to show in order to be able to secure new grants. *'Something to show?' Yes, at least three accepted papers in good journals. 'In four years??? '. If the work takes longer, this may not allow for these papers and you failed and will not be funded anymore. Career over!'* I often close this part of my talk a bit ironically: *'For him and me it worked well, I was first author, Melief, last author, he moved on to his next job and became a professor in 1986, I stayed behind and became a Fellow of the Royal Society and wrote my own grants and started my own lab on HIV/aids, was last author on the papers and became a professor in 1996. Science is as simple as that'.* Of course, the research style of Melief also in 1981 was not unique. True, he was an early adopter of the way biomedical research was to be done after *'the molecular turn'*. This was the real *Science in Action* (Latour, 1987) that Latour described which I in these days did read straight from the press. I must confess, I loved science from the very start. Some of the team and the department did not and still do not like it at all, as is also described by Latour (p155). They hated the need for networking and seeking allies, or to have to listen to the slick presentations of sometimes too weak data by competitive group leaders at meetings, the discussions with peer reviewers riding their hobby horses and other aspects of marketing and sales techniques. This was in their view embarrassing and even pathetic behaviour, more fit for short term politics, but surely not appropriate for the solid research they were doing at the bench, that had attracted them to a career in science.

committees populated by the elites of the various academic disciplines at any given time (Bourdieu, 2004; Polanyi, 1962). This is also how professional credits, reputation, academic positions and last but not least financial credit, research funds are distributed. This concept of a field and its credibility cycle was taken from the work of Bourdieu and visually depicted by Latour and Woolgar in their seminal study of the daily practice of knowledge production by biomedical scientists at the Salk Institute at San Diego (Latour & Woolgar, 1979).

3.8   Of High Church, Low Church

Over the years since the 1980s the system of science was increasingly held more accountable to its claims and promises on the return on investments. The external political causes relate to the growth of the system in researchers, the ever-increasing volume of investments required, the need for governments to make choices that could be explained and defended, based on data to show results in relation to societal and since 1990 dominantly economic needs. In that development, the life sciences and engineering were thriving, physics that did well in the Cold War with *user-inspired basic research,* suffered (Stokes, 1997). Increasingly also research on environmental sciences has been growing until now. As described (Rip & van der Meulen, 1996; Wouters, 2014), the national aim to compete with respect to the military during the Cold War and later mainly economically by investing in science, technology and development called for ways to quantitatively measure the impact of science. Since for societal impact to show a large lag time is required, short term quantitative measures were used mainly of publications and their impact via citations and numbers of patents. Gradually from 1980 the use of these metrics has become dominant to measure the performance of the system at the national and institutional levels and to the level of departments, laboratories and research groups.

Gradually since the 1980s it rapidly became normal practice in academia to use these indicators also for the evaluation of research of individual scientists. The nature of the indicators to choose were never discussed on beforehand in small committees or larger conferences and meetings. They evolved over the years and their use got established by the legendary 'invisible hand', being an interplay of concepts of science, and of interests and powers in the different academic communities as discussed in the previous chapters. Implicit and explicit ideas about hierarchies of journals had evolved and were linked to journal impact factors which became the measure, not only for the journal at large, but for the individual research paper published in the given journal. Not unexpectedly, in the natural, biological and biomedical sciences the idea of excellence became linked to a specific type of research inspired by the Legend. This modelled after the quantitative, formal and analytical type of the work done in physics. The emphasis like in the natural sciences was on more basic work resulting in general findings, and theories of a more abstract and theoretical type which was suitable for international English language journals with a broad readership. These journals by definition thus had a higher impact factor and they started to actively game this process in order to become the hottest journal for researchers to publish. They started, for example to solicit more reviews on topical issues and focussed on and invited 'sexy' research papers that presented novelty about hot topics that changed over time given developments in the field. The 'normal' solid science was rejected and advised to go to 'speciality journals' in my field for immunology, virology and infectious diseases for instance. In the same vein, qualitative scholarly work and applied research and papers reporting negative results became less valued and less easy to publish properly. This translated to a shift to reductionist formal methods of research also in other fields like economy, the geosciences, social-psychology, sociology, linguistics and even in the humanities. As 'high church' research scored in higher impact factor journals and was thus better regarded in career advancement or funding committees, it converted these academic credits in monetary (funding) credits which in turn were used to produce more of the required type of papers. At the higher organizational level, this type of academic output is important for the institution's position on international ranking lists.

Of note, this trend was coming from the gradually changing mores of the researchers serving on committees and via that route it became policy in committees at universities and funding agencies to use quantitative bibliometric indicators for quality that referred to internal academic excellence and not societal value and impact. This all relates to the accumulation of credit, scientific and social capital required for career advancement at the individual level and has resulted in an academic culture characterized by massive production of papers, a bibliometrics game driving for particular types of publications. Metrics are even changing how scientists define quality, relevance and originality in the first place (Müller & de Rijcke, 2017; Wouters, 1999). Production of robust and significant knowledge and results are secondary to short term output complying with a quantitative credit system for academic career advancement. This is primarily evaluated at the individual level which goes against collaboration and multidisciplinary team science in departments. There is, based on empirical data, wide consensus that this is the main factor that determines the semi-economical behaviour of researchers regarding problem choice, collaborations, networking, grantsmanship and publication strategies, funding and outreach (Bourdieu, 2004; Latour, 1987; Stephan, 1996). This highly competitive social system does result in a widely felt lack of alignment and shared value in the academic community (Fitzpatrick, 2019). These normative, opposing and often conflicting ideas about what science should be and the type of research excellent scientists should be doing, indeed still are the cause of many problems in academia. Within the field (of the social game) of science and research it has resulted in unsound competition, power struggles, elitism, stratification and hierarchy between academic fields and of note, within disciplines that are based on obsolete or simple wrong ideas about science and research. It has been shown by numerous studies now that across academia because of a massive growth of the numbers of scientists and investments, because of hyper specialisation social and quality control by institutional and peer review fails. This has led to a generally felt frustration by the majority of scientists in academia, which however the academic leadership did not immediately recognise, flatly denied or recognized but rebutted with '*this is how science is, if you cannot stand the heat get out of the kitchen*'. When it was acknowledged, then one was advised by mentors and colleagues not to address it openly, in order 'to not hurt one's own career chances in academia'. It in this way has had and still has major impact, in particular on the lives and careers of students, young and mid-career scientists.

Problems of the Current Reward System in Science

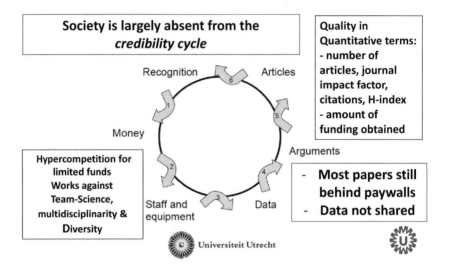

3.9   Physics Envy

It is clear that this system is not incentivising and rewarding investigators who do work in too close connection with ('messy') problems in the real world, as it does appreciate more fundamental ('pure) formal work in the natural sciences and bio-medicine, but also in the social sciences. The criteria and norms of excellence and concomitantly the dominant metrics being used to evaluate science and scientists across the institutions in academia were and are still strongly determined by the classical ideas about science, with the historical preference for the methodology and type of formal products of the natural sciences. This forms in academia a major well-known disadvantage to SSH compared to STEM and the biomedical sciences. In a response to survive and compete, researchers in social sciences, economics and even humanities in the past 20 years took refuge to research with more quantitative methods aiming for more general theories and insights. This **'physics envy'** serves to show that their methods and conclusions are 'hard' science as well. As a conse-quence in these academic disciplines, including the biomedical sciences, there developed a visible gradient from quantitative physics-like research to classical scholarly humanities work not using math but reasoning and argumentation. This is the gradient from 'high church to 'low church' as I call it.

Playing the Games of the Field (3)
Getting his attention!

On a spring morning in 1991, Hanneke Schuitemaker and I had, after heavy negotiation with his secretariat, an 8.00 am meeting scheduled for just 15 min. Knowing that he started in the office at 6 am and worked till very late and mostly on formal more important dossiers, we were prepared. I had adopted that practice and nearly every year visited NIH to discuss with the important researchers at NIAID, then and still (!) lead by Dr. Anthony (Tony) Fauci and his collaborators. Fauci, now 79 of age, is still very much in that job, now in daily White House press briefings because of the COVID-19 pandemic. He was already then busy and extremely efficient with his time, always 1 day 'in and out' of conferences giving his famous ultra-speed keynote talks on data from his own laboratory. We were that morning going to show Fauci unpublished work from Hanneke with evidence for two different strains of HIV with pathological and clinical implications. We knew we had to talk for 15 minutes straight without a pause to inhale, because we feared that Fauci would otherwise takeover and start to tell us about his work. It was a rehearsed marketing and sales pitch for our SI and NSI viral phenotypes. We apparently succeeded. Many years later after the molecular confirmation and identification of their receptors from many labs, Fauci at a meeting referred to our pitch. So the advice is: 'spread the news about your 'important' work at visits around the globe, at the corridors during coffee breaks of the meetings, ski lifts, and especially at the 'gossip sessions' at the bar during meetings, at speaker dinners and of course on TV shows if you get a chance. You never know for which major journals they are a reviewer or on which committees these people might serve. I hasten here to give Fauci the credit he deserves for his current role in dealing with the COVID-19 pandemic in the US, but in the context of the early days of the aids pandemic for engaging with the gay community in New York and truly listening to their complaints and needs. Fauci has been more or less personally responsible for the formidable budgets coming to NIH to fight HIV and AIDS. In those early years when the US government was not that receptive, the gay community in their frustration unfairly put the blame on Fauci, but soon recognized him to be a loyal partner in the fight against HIV/AIDS.

(continued)

Photo: National Institutes of Health Library

3.10   Science in Transition: The Initial Reception

Before the official international start a symposium was planned on 7 and 8 November, but we first organized a small format on-invitation meeting to get a first response to a near final draft of the Position Paper and commitment from the field on September 25, 2013. We had invited representatives of the various players in the domain of science and society. These included the association of universities in Netherlands (VSNU), The Royal Academy (KNAW), governmental funder the Dutch Science Council (NWO/ZonMw), the representative of the joined federation of Dutch charities and directors of intermediate institutes that advice the government on science, innovation and development. The latter included the Netherlands Scientific Council for Government Policy (WRR), the Netherlands Environmental Assessment Agency (PBL) and The Rathenau Institute. The reactions were, as anticipated, quite mixed. Some, especially the representatives from the Royal Society, the universities and The Dutch Science Council felt that the tone was harsh and suggestive of a crisis for which data, they thought, were lacking since mostly anecdotal stories were reported. Some felt offended, and even doubted that something was wrong at all. In general though, the fact that by our position paper this debate is now in the open was appreciated, although fear for backfire from politics and society was abundant in the group. It was believed that more empirical evidence was needed to better estimate the size of the various problems and get a feel for the international and historical perspectives. It was agreed that the relation between research and teaching and the interaction with society, needed more attention. Finally, it was felt that given the issues that were brought up, the adverse effects of critical parts of 'the system' needed more research. Bert van der Zwaan, from geosciences and then the Rector of Utrecht University, after being critical and irritated about the logic and unpolite tone of our paper, was clearly in agreement with our idea that actions should be undertaken to change the incentive and rewards system. Hans Clevers, an internationaly well-known researcher in biomedicine, then the President of the Royal Academy said that he, as an active researcher in stem cell and cancer biology, recognized the issues and was sympathetic to the proposed actions to be undertaken.

Rutger Bregman, historian and journalist at *De Correspondent* announced to start to practice investigative journalism into science, in analogy of how Joris Luyendijk researched the financial industry of the London City, to find out about a crisis of the system. This idea of Bregman was repeated by me as an invitation to the participants of a meeting of the Dutch science journalists held in October at the Royal Academy. In the weeks before the symposium of 7 and 8 November, Volkskrant a major national newspaper announced an investigative series on how science really works. The Utrecht University journal DUB started a science blog around the Science in Transition debate. *Economist* came in October with an impressive well-researched issue on 'How science goes wrong. *Scientific research has changed the world. Now it needs to change itself'* https://www.economist.com/leaders/2013/10/21/how-science-goes-wrong.

The articles in *Economist*, much to our surprise, followed largely the main criticisms of our position paper with evidence. Our response was, 'hey, they stole our thunder', but we were pleased also, because those who questioned our analyses and called for more evidence were being served. Had we only known how much more of that evidence was to come in the next few years! Already in the days immediately before Thursday 7 November there was media coverage. Saturday, 2 November NRC Wetenschap had a very constructive main article about Science in Transition. Hendrik Spiering, Chief Editor of Science News/Wetenschap of NRC who was a columnist on Friday, had written a main Editorial in the newspaper of Wednesday 6 November, on Science in Transition. The morning of 7 November, Volkskrant featured a large interview about the Science in Transition. Where I frankly explained the perverse incentives and argued for a more socially responsible research agenda to make research more relevant for society. Next to DUB and Folia, the magazines of the University of Utrecht and Amsterdam announced the symposium. As a surprise at breakfast, the Saturday morning after the meeting Volkskrant had a large piece with a figure showing the credit cycle! This was based on a slide I had started to use in these years and still use, adapted from Laurence Hessels (Hessels et al., 2009). Each day the symposium was attended by approximately 200 people. It was in the subsequent days and week covered in many newspapers and radio interviews. The evening of 7 November, I was in a nine-minute live interview on *Nieuwsuur*, a high-quality late-night news program. Some in the science community were absolutely not amused by the tone and style of how we presented our conclusions and our case for change. *'Not so much that there are no issues, but research is by far not as grim as your story suggest, and this is going to undermine trust and is going to decrease funding from government.'*

Prof. Jan Vandenbroucke, in an exchange earlier that year, disproved of the contrast of the Legend of science, the positivistic idea of the objective 'scientific method' and its Mertonian norms (Chap. 2), with the less romantic social reality of how knowledge is produced in the workplaces of science and research. He argued that both are part of the more realistic practice of science and 'fierce competition and jealousy' do not inhibit or interfere with the growth of knowledge. It is, he says, exactly criticism and strong debates that are needed to arrive at reliable knowledge. He cites Stephen Gould who in the context of the Science Wars has argued that these views of research can be understood to be parts of our daily research practice and that this is the social way in which we produce 'objective' -or did Gould mean 'intersubjective'- knowledge that we accept as 'truth? With this I agree. I have argued in Science 3.0 and the Position Paper, that once we leave the positivistic Legend behind, we can explain in honesty as Gould does, how we arrive at accepted claims that are not absolute timeless truths but always subject to tests and criticism. So, where is the problem between Vandenbroucke and us? In an email to me in response to the Position Paper in September 2013 and in follow up of the debate, Vandenbroucke clarified the issue. He does not, as I do, believe that the positivistic Legend has a deforming effect on the practice of science. I am fighting a ghost he says.

3.11   Science in Transition on Tour

After the symposium, Jerome Ravetz left the scene, he had done his job and found it too difficult to participate any longer from his home in Oxford. We were invited to organize an afternoon session on *Science in Transition* at the 2013 WTMC annual meeting, on November 29. Huub Dijsterbloem, Frank Miedema, Paul Wouters and Hans Radder presented, for an at least for me, quite intimidating audience of scholars, including the members of the WTMC International Advisory Board, Aant Elzinga, Tom Gieryn, Steven Shapin and Andrew Webster. My point to them was: *'You have been studying and writing about science and its institutions. STS has over the past 30 years obtained the status of a well-respected discipline in SSH and academia. Now it is time 'to translate this 'pre-clinical' knowledge to the 'clinic' were the patients are. We have a problem in university, and we need you and your knowledge badly.'*

The Dutch initiators received and accepted many invitations to present and explain the message of Science in Transition at universities in the country. On our website we had the agenda with these activities to show to interested people the reception and that the movement was alive. In 2014, virtually at every university and academic medical centre one of us presented and debated. In these days the audience recognized the issues and urged us to present more of the interventions needed. The Boards of universities, we were told at some of these meetings, were not all amused, they feared it could cause unrest. Particularly with regard to the use of metrics, it obstructed with all institutes heavily playing the Shanghai Ranking. I here must be honest, since I as researcher, professor and institutional administrator, also had until very recently been 'addicted to the Journal Impact Factor'. A confession I still often use to start my seminars with. It must be said that the rectors of University of Amsterdam (UvA) and Leiden University in January and February in their Dies speeches supported the initiative. De Jonge Academie of the KNAW in February presented a Vision on science and research that echoed many of the issues.

Science in Transition Conference: November 7 and 8, 2013, KNAW Amsterdam

Over the next few years, science will have to make a number of important transitions. There is a deeply felt uncertainty and discontent on a number of aspects of the scientific system: the tools measuring scientific output, the publish-or-perish culture, the level of academic teaching, the scarcity of career opportunities for young scholars, the impact of science on policy, and the relationship between science, society and industry.

The checks and balances of our scientific system are in need of revision. To accomplish this, science should be evaluated on the basis of its added value to society. The public should be given a better insight in the process of knowledge production: what parties play a role and what issues are at stake? Stakeholders from society should become more involved in this process and have a bigger say in the allocation of research funding. This is the view of the Science in Transition initiators Huub Dijstelbloem (WRR/UvA), Frank Huisman (UU/UM), Frank Miedema (UMC Utrecht), Jerry Ravetz (Oxford) and Wijnand Mijnhardt (Descartes Centre, UU).

Location: Tinbergenzaal, KNAW Trippenhuis, Kloveniersburgwal 29, Amsterdam.

Key notes by Sheila Jasanoff (Pforzheimer Professor of Science and Technology Studies, Harvard Kennedy School) and Mark Brown (Professor in the Department of Government at California State University, Sacramento); Column: Hendrik Spiering (Chef Wetenschap/ Editor NRC Science): *Nieuwe tijden, nieuwe wetenschap*

(continued)

Speakers: Sally Wyatt (Professor of Digital Cultures in Development, Department Technology and Society Studies, Maastricht University); Henk van Houten (General Manager Philips Research); Hans Altevogt (Greenpeace); Jeroen Geurts (Chairman Young Academy KNAW, Professor Translational Neuroscience VU Medical Center); Rudolf van Olden (Director Medical & Regulatory Glaxo Smith Kline Netherlands); Peter Blom (CEO Triodos Bank); Jasper van Dijk (Member of Parliament Socialist Party); Hans Clevers (President of the Royal Netherlands Academy of Arts and Sciences (KNAW). Panel discussion with: Jos Engelen (Chairman Netherlands Organisation for Scientific Research (NWO); André Knottnerus (Chairman Scientific Council for Government Policy (WRR))Lodi Nauta (Dean Faculty of Philosophy, Professor in History of Philosophy, University of Groningen); Wijnand Mijnhardt (Director Descartes Centre for the History and Philosophy of the Sciences and the Humanities/Professor Comparative History of the Sciences and the Humanities, Utrecht University)

Folia, the weekly of the University of Amsterdam featured Dijstelbloem and me in a discussion with UvA professors who were quite critical.

We were invited for a discussion with Jet Bussemaker, the Minister of Higher Education who was very interested. We discussed at the Royal Society with Directors of the KNAW Institutes where we were met with support, interesting suggestions for improvement and heard the familiar objections: that 'if we engage the public they will not allow for basic science and novel programmes', that they don't understand science, and of course from the natural sciences *'When I am hiring, I judge scientists on the JIF of their publications. If that is abandoned, what shall we use instead? Anyhow, it will take much more time.'* We tried with: *'.....uhhh, just an idea, whar about reading their selected papers?'*

We met with the Board of NWO, the major Dutch government funder board, who were really not amused at all. In a meeting with the chair and director of the Association of Universities in the Netherlands (VSNU), who were much more engaged already we discussed the effects of the current Incentive and rewards system. We pitched at the 'Night of Science' of UvA and Hans Clevers in his annual speech as President of KNAW discussed some of the hot topics. In June 2014 we published our evaluation of an academic year of Science in Transition and announced we would continue, because of enormous support and because we were even more convinced of urgency and need.

In the summer of 2014 the European Commission, the Directorate-General for Research and Innovation (RTD) and DG Communications Networks, Content and Technology (CONNECT) started a public consultation under the heading *'Science*

2.0': Science in Transition. The accompanying background document written by René von Schomberg and Jean Claude Burgelman presents an analysis of the current state of science and how science could change to be more efficient and may contribute more to society (EU, 2014). In a section called Science in Transition a few ongoing initiatives driving for change are discussed. Many of the issues are in agreement with the *Science in Transition* analysis and the authors state: *'In the Netherlands, an intensive debate has evolved on the basis of a position-paper entitled 'Science in Transition'. The ongoing debate in the Netherlands addressed, among other, the issue of the use of bibliometrics in relation to the determination of scientific careers. However, this debate went actually beyond the scope of what is described in this consultation paper as 'Science 2.0' and included also discussions on the democratisation of the research agenda, the science-policy interface and calls for making research more socially relevant.* This questionnaire and the very informative analysis of the results were the start of the EU Open Science program in 2015. It appeared that many stakeholders preferred 'Open Science', not only as an alternative term over 'Science 2.0' but more importantly they liked to see science make the transition to the practice of Open Science. This policy transition to Open Science by the EU, in my mind was critical and will be discussed in more detail in Chap. 7.

A presentation on Science in Transition was given in September in Brussels for the policy advisors of *Science Europe*, the European association of public research performing and research funding organisations. One of them said that she like the ideas and plans a lot, but *'did I know why the ERC was established next to FP7 and Horizon 2020? To serve those who want to get ample funds to do 'free curiosity-driven research and not be bothered.'*

The Dutch Ministry of Higher Education, Culture and Research, with reference to the debate elicited by Science in Transition organized debates to prepare for an integral vision and mission of research and science for the new government. Their

---

The Elephant in the University Board Room

'It was a bright and sunny afternoon in June 2014, when members of the *Science in Transition* team met with the Rectors of the Dutch Universities at Utrecht University's *Academiegebouw*. The meeting took place 7 months after the first symposium, which had inspired a national discussion about the state of the art of in science and academia. The message of Science and Transition was initially met with a lot of sympathy by those who recognized the problems and their potential causes. Many liked the interventions suggested by Science and Transition to improve science and academia. But some complained about the polemical way the message had been delivered in the media. While they agreed with the analysis, they were afraid that it might backfire on science and scientists.

(continued)

Others said the analyses were not new at all, as they were being discussed for years already. Lastly, there were those who rejected the analyses of SiT altogether, arguing that there was no need to change: science is an international endeavour, and the Netherlands were doing an excellent job in the rankings. All of these criticisms were aired that Thursday in June during the first 30 min of our meeting. Then the Rector of the University of Amsterdam, Dymph van den Boom intervened. She stopped the discussion and said: *'Dear colleagues, let's face it, there is a big elephant in the room. It may not have been particularly nice how our guests talked about our science and our universities, but they definitely have a point'*. That started the conversation.'

In some respect the Rectors have to be excused for their slow response. Just before our public debate in 2013, Hans Radder, who had been engaged with us, had with Willem Haffman published an Academic Manifesto which put all the blame on the university administrators (Halffman & Radder, 2015). They had sold academia due to the neoliberal evil of private interests, driving for patents (patenting they believed should be abandoned anyway) and financial gains. They had turned scientists into capitalist entrepreneurs instead of working for the public good. It may be that the Rectors also regarding Science in Transition sensed that something much worse was in the air. Indeed, 9 months later in Amsterdam a far more radical and uncontrolled up rise started in the University of Amsterdam with squatting of the Maagdenhuis, the home of the University Board which resulted in the stepping down of the Board. This movement called *Re-Think* was more in line with Radder and Haffman's Manifesto, many complaints and a call for

(continued)

academic autonomy and for the democratization of university government and in a sense arguing for insulation from influences from society. In their eyes we, Science in Transition, were not to be trusted because too close to the people in power in academia. In our eyes they were not forward looking and did not present a clear integrated vision on science and academia in the twenty-first century.

Science Vision was proudly presented in November 2014. December 3, the second Symposium was held at KNAW about transitions, with international and national discussants. At that occasion the Association of Dutch Universities signed DORA (for the first time).

3.12 Metrics Shapes Science

The style of the *'high church'* of research remained the style of research with the highest esteem in academia and public research institutes. Accordingly, a credibility cycle with indicators derived from that type of esteem and excellence was dominantly used in distribution of reputation and funds in heavy competition by classical peer review schemes. This is reflected in the appreciation of pure/basic over applied science, formal quantitative (modern) over qualitative and argumentative research. Also think of the scientific status of the 'hard' sciences over the 'soft' sciences and correspondingly the potential impact of investments in natural and biomedical sciences over those in humanities and the social sciences. This system with its dominant indicators thus has major effects on the agenda setting of our research. Since these problems have been put forward by a now increasing number of writers from within academia, the issues are also increasingly experienced by administrators in university, funding agencies, government and elite key opinion leaders in academia. In reaction to that conservative view and reward system from academia, alternative institutes and funding schemes were developed, initially mainly by governments to accommodate mission-driven science for which next to scientific excellence, quality criteria related to reliability, robustness in practice and thus to societal impact was important. Here, researchers do work in national and international teams and consortia on complex real-world problems, many times in collaboration with private partners and citizens involved. This was and still is to a large extent by the academic elites regarded as *'low church'* research because it is done with less competitive, soft money and thus these types of grants, such as those from FP7 or Horizon 2020, come with much less esteem that a grant from the ERC. This is just the old academic elitist game being played over and over and here on the distinction between pure and applied science and on winning in competition. It needs no explanation

Level Playing Field? (4)

The popular image of science, as we saw in Chap. 2, is based on a community of researchers with, if not unique, for sure, exceptional integrity and altruism. They follow their professional vocation to search for truth and do this openly, disinterestedly and with great unselfish honesty. It was admitted by Merton, there is the Matthew Effect and inequality and there are elites. It was believed that especially the top scientists are endowed with exceptional integrity to serve as role models for those who are in the heat of the daily competition. Advancing in the field, scientists realize there is more stake than finding significant insights and knowledge. It is very much about who first discovered an insight. Moreover, major novel insights are threatening as they overthrow major previous results of leaders in the field and are generally resisted and not immediately accepted. When you are not generally seen as a major player, work has to be done to make the community aware of an interesting result and get the credits badly needed to survive in the system. During the first years as a group leader I learned some 'tricks of the trade', pushing the findings of your laboratory, which after reading Jim Watson's *The Double Helix* were not that surprising anymore.

In 1987 in a collaboration with Hidde Ploegh and his colleagues, then at the Netherlands Cancer Institute, we observed that by inhibiting enzymes that are important for the sugar coating of the HIV envelope protein the interaction with the receptor on human T cells was disturbed. HIV was rendered non-infectious. This was biochemically of interest and opened up avenues for anti-viral drug development. Hidde was the major and thus last author and decided 'to go for Nature'. The review reports, at that time by airmail, were not all that favourable. No problem for Hidde who had at that time already broad international experience and standing in the field as a top biochemist and immunologist. In my presence he simply called the editor, they discussed the comments and Hidde explained why the thought not all reviewers appreciated the significance of the work. A fourth expert was asked to review and November 5, the day after my oldest son was born the paper was published and was prominently featured in The Volkskrant, a respected national newspaper (Gruters et al., 1987).

Nine years later, in January 1995 two major, very innovative papers were published in *Nature* that shed new light on the dynamics of HIV infection and urged us to rethink the immunopathogenesis of AIDS (Ho et al., 1995; Wei et al., 1995). The authors were interviewed on CCN and made headlines in major newspapers around the world. We had been engaged in experiments to test the old hypothesis and came to the conclusion that the old hypothesis was wrong, but our data also provided unexpected amazing evidence against the major immunological component of the new hypothesis proposed by Ho et al. As David Ho then was one of the major scientist in the field, I thus anticipated

(continued)

resistance from reviewers to our data and decided to make a bold action. In a rooftop restaurant overlooking the harbour of Vancouver, at the occasion of the XIth International AIDS Conference in Vancouver in July 1996, I met with an editor of *Science*. At the meeting, the new hypothesis was the hottest topic by far, with in the meantime new papers by these same authors in major journals.

Over dinner I explained our data and its implications in detail. She was very interested and after the desert and coffee, asked me to submit as soon as possible. As anticipated the reviewers thought the data, intriguing, but they were not sure and in the end found the data hard to believe. 'Because', one said, 'if this is true then even the new immunology hypothesis is not correct'. The paper was improved by taking these comments into account and was published in Science in November 1996 (Wolthers et al., 1996). Fortunately, our data were confirmed very soon.

You think I was addicted to the JIF? Yes, I was, because we knew that papers in these journals were regarded very important and instrumental to convince the community and our peers in the national review boards of our findings. They also definitely helped me to get my appointment as professor that same year. I hope that for experts, it was not the JIF, but our data that made the difference. Speaking about impact, David Ho, the major principal investigator and advocate of the new hypothesis of the *Nature* papers was elected Man of the Year of 1996 by *Time Magazine*.

that research done with whatever type of money of course can result in excellent research in its own right.

3.13 It Is Contagious?

Could this view and this practice of science, the reader might secretly hope, not be a 'Dutch Disease', driven by dangerous liaison between Calvinism and neoliberal capitalism? The answer, I am afraid, is a clear no. This system of incentive and rewards, informed by the Legend and its legacy of the myth of the scientific method of reductionism, has shown to be highly contagious and has been disseminated as an infectious disease by academics travelling all over the globe. It has in the past 20 years become common practice in Europe, Canada, Australia, India, Indonesia, China, Singapore and Hong Kong, Latin America and in Sub-Saharan Africa, most notably South Africa. The introduction of international rankings, especially the Shanghai Ranking in 2006 has accelerated the use and abuse of the metrics in the

Distortions of the Practice of Science and Research
STEM dominate over Social Science and Humanities
    Theoretical & pure science dominate over applied science and technology
    Curiosity-driven research is believed the best for solving societal problems
    Scientific knowledge is neutral and value free and science should be autonomous, not bothered by external publics or politics and their problems. Scientists cannot be held responsible for the knowledge they door do not produce
    Quantity, Replication, Relevance and Impact are subordinate to *novelty and quantity*
    Individual Hyper-competition works against Team-Science, Multidisciplinarity and Diversity
    Universities outsource talent management to funders based on flawed metrics, instead of having a research strategy according to their mission
    Short-termism and risk aversion is rife because of four-year funding and evaluation cycles
    Fields with high societal impact, but low impact in the metrics system suffer (aplied vs basic; local vs international)
    The national and institutional research agenda is not properly reflecting societal (clinical) needs and disease burden
    Open Science research practices are just 'nice to have': stakeholder engagement, FAIR DATA, Open Code and Open Access

Who Sets the Research Agenda of the Field? (5)
When in 1981 the first aids cases presented in the US and later all over the world, it was quickly understood that an infectious agent, most likely a virus was the cause. It was sexually transmitted by body fluids, like blood and thus also blood products, for which good evidence was produced early on. Patients presented and died because of compromised immunity and it soon appeared to be associated with a loss of a specific population of white blood cells, so called helper CD4 T cells. At CLB, one of the predecessors of Sanquin, the Dutch Blood Supply Foundation, the new virus was a serious threat to the safety of the blood supply and called for immediate action. Virology at that time, was not a big thing. In times of COVID-19, knowing now, what has happened since 1980, with HIV, SARS and Ebola and major Flu pandemics, that is hard to believe. At that time, it was thought that we had won the war against viruses and not much academic reputation and funding was to be obtained in

(continued)

human virology. There was, driven by medical microbiologists an effort on Hepatitis B Virus and to some respect on non-A- non- B Hepatitis Virus, which later was called Hepatitis C Virus. Medical microbiology was a very applied art, important for patient care and public health but academically regarded a done job. Identifying new viruses, for instance in seals, what our now famous colleague Ab Osterhaus at that time was doing, was pitied by scientists and compared to 'collecting rare stamps'.

The Melief lab where I was, was involved in tumour-immunology in murine models. Given the career of Melief, an MD who was raised in the setting of blood transfusion and blood products, he was open to go to human research. He studied the development of murine leukaemia caused by mouse retroviruses, following the then widely held belief that viruses caused cancers also in humans. In the past 40 years there is much more evidence for that, but at that time this had been shown for Epstein Barr virus causing Burkitt's Lymphoma and chronic Hepatitis C Virus infection associated with liver cancer. Retroviruses, related to those known to cause tumors in mouse and cats were sought in humans but not known this changed in 1980 when the first bonafide novel human retrovirus was identified by NIH researchers lead by Bob Gallo and a group in Japan lead by Hinuma. This virus caused a rare cancer of white blood cells prevalent in the population in Japan and the Caribbean. It happened that my project was on human T-cell leukaemia and Melief started a collaboration with colleagues, who treated leukaemia patients in the Caribbean communities in Amsterdam and London to study the involvement of the virus. I brought tests detecting immune responses to HTLV-1 to the lab from London and indeed found evidence for the presence of the virus in T-cell leukaemia patients. In 1982 when the first aids patients also presented in The Netherlands, there appeared a claim in the literature that HTLV-1 might be involved. We started a collaboration with Jaap Goudsmit a medical microbiologist and virologist at AMC, who was keen to find an interesting and challenging new research topic and had spotted aids as an ideal candidate. We tested whether evidence could be found for HTLV-1 infections in AIDS patients in Amsterdam. There was no convincing evidence, but my career had made a dramatic turn to research on HIV/aids already. From then, I worked on viro-immunology of HIV and aids which was driven by the urgent problem that HIV caused for the safety of various blood products. The murine virology at the institute had stopped in 1984 as Melief had followed the field to work on oncogenes in mouse and humans. Oncogenes had just then been discovered in models of murine and Rous sarcoma tumour viruses, research that was propelled by enormous technical progress in molecular biology in the late 1970s. So, in 1984 Melief and his group logically left for the Netherlands Cancer Institute.

incentive and rewards system. As we have seen this put most of the weight on the more basic science, and the publication and citation cultures of STEM. Science nowadays must be 'international' to score, work on urgent national and regional problems normally does not get published in the English language top journals. The effect is that in order to get higher in the rankings, research in universities in for instance Indonesia or South Africa is steered towards topics that score in international high impact journals, at the expense of research on topical problems and needs of the local publics. I don't even mention that most institutions in developing countries cannot afford the subscription fees of the top journals, most which are not open access. Results of our HIV/aids research done in Amsterdam were not accessible to researchers and medical specialists in developing countries that had the greatest disease burden with social disruption and literally millions of deaths from aids. Only in an acute crisis as the COVID-19 pandemic that we experience at this time, all data and papers are made immediately open and accessible to all. Will this openness be only temporarily?

3.14   Interventions Needed

Most of the different components of the analysis of Science in Transition, as out lined above, at that time were not new or original at all. I can cite many more well-written and well-documented texts, in journal articles and books that analytically tells us the same. A fine example is European Science Foundation Science Policy Briefing, written at the same time as our position paper (ESF, 2013) in 2012/2013 by seven top experts amongst others Ulrike Felt, Alan Irwin and Arie Rip. The paper explicitly discusses the adverse effects of metrics on problem choice and the fact that public engagement is not being considered as part of the research and suggests interventions, as DORA (p20–21). The authors however did not take the next step to list a series of concrete actions to be taken by administrators and scientists who are responsible for that problematic and limited system of research evaluation. I have pointed out that in general most of these authors stayed at safe distance from the proverbial elephant in the room.

Discussions about changes in this part of the governance system of science and research have intensified in recent years. It was pretty normal in the 1960s and 1970s to talk about science in terms of power, elites and money. Since the 1990s that talk seemingly was taboo. It seems to me that since 2015 or so the taboo has been broken, hopefully for good. We needed to open the black box, of how science as an industry is being run and by whom, to expose and make visible the machinery of what the classical sociologist of science called 'the invisible hand'. Like in an unregulated economy, the invisible hand, not unexpectedly, when made visible appears to belong to the powerful and the elites of the day. In this case thus very much the scientists who did well in the social system described above. They strongly believe in the Legend and its metrics, some honestly and for real, but others used it as a masquerade, a 'front stage' mythical image that still sells science well to public and politics. We have seen that the myth is scientifically but also socially untenable. In 'modernity', that is to say in our modern times, the public in the new social media is in uncompromising open interaction with science in its many forms. On that boundary of the science of complex societal problems and society, there is no consensus, no absolute truth and the public increasingly gets to see more of the backstage practice of science, where the discussion have not settled but are raging as they always did. In these reflexive times in society, we need a more reflexive narrative about how we do, and with whom, and for whom we do science and research. There are as we see in the next chapter many small-scale ongoing movements and actions to build this reflexive narrative. In many of these this is done together with people from outside academia that have a stake in the research because it is their problem that is to be investigated. In these transitions there is awareness that the publics will talk back. We need to let go the idea of *'The Quest for Certainty'* and relate to *'The Public and its Problems'* in order to produce not absolute truths, but significant reliable knowledge that benefits us all.

3.15   Sensing the Zeitgeist

During the Christmas break of 2014, reflecting on the start and the reception of the message of Science in Transition in the first year in The Netherlands, we were surprised and amazed. We had expected some reactions and a bit of media attention when we prepared the paper and the symposium. We had not anticipated the enormous and sustained support, from academia and outside academia nor the media attention and exposure. What we had expected was a typical half-hearted response from the leadership, with a standard reflex that this 'was all already known and adequately addressed' by the Boards and Deans. After that, we thought our message would for surely fade away quickly replaced by other news. We even had been prepared for straightforward denial and rejection by the establishment. Some of these reflexes were heard and seen in writing. The response generally however was positive from many different corners and echelons inside and outside academia. Our analysis was widely recognized and brought palpable relief that it was now acceptable to openly discuss these issues without being scorned as a complaining loser. In addition, the debates did not stop with pointing out the problems but included actions and interventions at the systemic level.

This description of the reception is provided here. Not to show how unique or enormously clever we were, because in fact we weren't, as some colleagues were happy to point out. It is to illustrate the widespread criticism, critical insights and frustration that became tangible and had apparently been building up in academia over the years. Obviously, this was not the effect of our initiative. It was already in the air, after years of critical thinking and writing by many colleagues in different countries. In addition, it was fuelled by increasing massification and digitalization, by the distorting effects of the neoliberal knowledge economy and its New Public Management. This somehow had been brewing for a decade and the science community was ready for this broad and international call for change. It was this *Zeitgeist* that had activated us to take action, to give, like others had been doing at the same time elsewhere, a small push.

References

Altman, D. G. (1994). The scandal of poor medical research. *BMJ, 308*(6924), 283–284. https://doi.org/10.1136/bmj.308.6924.283

Begley, C. G., & Ellis, L. M. (2012). Drug development: Raise standards for preclinical cancer research. *Nature, 483*(7391), 531–533. https://doi.org/10.1038/483531a

Bourdieu, P. (1975). The specificity of the scientific field and the social conditions of the progress of reason. *Information (International Social Science Council), 14*(6), 19–47. https://doi.org/10.1177/053901847501400602

Bourdieu, P. (1988). *Homo academicus*. Stanford University Press.

Bourdieu, P. (2004). *Science of science and reflexivity*. Polity.

Bourdieu, P. (2010). *Distinction: A social critique of the judgement of taste*. Routledge.

Butler. (2007). Assessing university research: A plea for a balanced approach. *Science and Public Policy, 34*(8), 565–574.

ESF. (2013). *Science in society: Caring for our futures in turbulent times*. Retrieved from http://archives.esf.org/fileadmin/links/Social/Publications/spb50_ScienceInSociety.pdf

EU. (2014). *Background document. Public consultation 'science 2.0': Science in transition*. https://www.researchgate.net/publication/275342087_Validation_of_the_results_of_the_public_consultation_on_Science_20_Science_in_Transition

Fitzpatrick, K. (2019). *Generous thinking: A radical approach to saving the university*. Johns Hopkins University Press.

Funtowicz, S. O., & Ravetz, J. R. (1993). Science for the post-normal age. *Futures, 25*(7), 739–755. https://doi.org/10.1016/0016-3287(93)90022-L

Gruters, R. A., Neefjes, J. J., Tersmette, M., de Goede, R. E., Tulp, A., Huisman, H. G., … Ploegh, H. L. (1987). Interference with HIV-induced syncytium formation and viral infectivity by inhibitors of trimming glucosidase. *Nature, 330*(6143), 74–77. https://doi.org/10.1038/330074a0

Guston, D. H. (2000). *Between politics and science: Assuring the integrity and productivity of research*. University of Chicago Press.

Halffman, W., & Radder, H. (2015). The academic manifesto: From an occupied to a public university. *Minerva, 53*(2), 165–187. https://doi.org/10.1007/s11024-015-9270-9

Hammarfelt, B., & de Rijcke, S. (2014). Accountability in context: Effects of research evaluation systems on publication practices, disciplinary norms, and individual working routines in the faculty of Arts at Uppsala University. *Research Evaluation, 24*(1), 63–77. https://doi.org/10.1093/reseval/rvu029

Hazelkorn, E. (2011). *Rankings and the reshaping of higher education the battle for world-class excellence* (pp. ix, 259 p). Retrieved from Ebook central http://ebookcentral.proquest.com/lib/oxford/detail.action?docID=678802

Hessels, L. K., van Lente, H., & Smits, R. (2009). In search of relevance: The changing contract between science and society. *Science and Public Policy, 36*(5), 387–401. https://doi.org/10.3152/030234209x442034

Hessels, L. K., Lente, H. v., Smits, R. E. H. M., & Grin, J. (2011). Changing struggles for relevance in eight fields of natural science. *Industry & Higher Education, 25*(5), 347–358. https://doi.org/10.1874/225580

Hicks, D., Wouters, P., Waltman, L., de Rijcke, S., & Rafols, I. (2015). Bibliometrics: The Leiden Manifesto for research metrics. *Nature, 520*(7548), 429–431. https://doi.org/10.1038/520429a

Hicks.D. (2012). Performance-based university research funding systems. *Research Policy, 41*(2), 251–261.

Hirsch, J. E. (2005). An index to quantify an individual's scientific research output. *Proceedings of the National Academy of Sciences of the United States of America, 102*(46), 16569–16572.

Ho, D. D., Neumann, A. U., Perelson, A. S., Chen, W., Leonard, J. M., & Markowitz, M. (1995). Rapid turnover of plasma virions and CD4 lymphocytes in HIV-1 infection. *Nature, 373*(6510), 123–126. https://doi.org/10.1038/373123a0

Ioannidis, J. P. (2005). Why most published research findings are false. *PLoS Medicine, 2*(8), e124. https://doi.org/10.1371/journal.pmed.0020124

Ioannidis, J. P., Nosek, B., & Iorns, E. (2012). Reproducibility concerns. *Nature Medicine, 18*(12), 1736–1737. https://doi.org/10.1038/nm.3020

KNAW. (2012). *Responsible research data management and the prevention of scientific misconduct.* Retrieved from https://www.knaw.nl/nl/actueel/publicaties/responsible-research-data-management-and-the-prevention-of-scientific-misconduct

KNAW. (2013). *Vertrouwen in wetenschap.* Retrieved from https://www.knaw.nl/nl/actueel/publicaties/vertrouwen-in-wetenschap

Latour, B. (1987). *Science in action: How to follow scientists and engineers through society.* Harvard University Press.

Latour, B., & Woolgar, S. (1979). *Laboratory life: The social construction of scientific facts.* Sage Publications.

Laudel, G., & Glaser, J. (2006). Tensions between evaluations and communication practices. *Journal of Higher Education Policy and Management, 28,* 289–295. https://doi.org/10.1080/13600800600980130

Medawar, P. B. (1982). *Pluto's republic: Incorporating the art of the soluble and induction and intuition in scientific thought.* Oxford University Press.

Miedema, F. (2012). *Science 3.0. Real science real knowledge.* Amsterdam University Press.

Moffitt, T. E., Arseneault, L., Belsky, D., Dickson, N., Hancox, R. J., Harrington, H., ... Caspi, A. (2011). A gradient of childhood self-control predicts health, wealth, and public safety. *Proceedings of the National Academy of Sciences of the United States of America, 108*(7), 2693–2698. https://doi.org/10.1073/pnas.1010076108

Moore, S., Neylon, C., Paul Eve, M., Paul O'Donnell, D., & Pattinson, D. (2017). "Excellence R Us": University research and the fetishisation of excellence. *Palgrave Communications, 3*(1), 16105. https://doi.org/10.1057/palcomms.2016.105

Müller, R., & de Rijcke, S. (2017). Thinking with indicators. Exploring the epistemic impacts of academic performance indicators in the life sciences. *Research Evaluation, 26*(3), 157–168.

Nagel, E. (1961). *The structure of science; problems in the logic of scientific explanation.* Harcourt.

Nosek, B. A., Spies, J. R., & Motyl, M. (2012). Scientific Utopia: II. Restructuring incentives and practices to promote truth over Publishability. *Perspectives on Psychological Science, 7*(6), 615–631. https://doi.org/10.1177/1745691612459058

Polanyi, M. (1962). The republic of science. *Minerva, 1*(1), 54–73. https://doi.org/10.1007/BF01101453

Prinz, F., Schlange, T., & Asadullah, K. (2011). Believe it or not: How much can we rely on published data on potential drug targets? *Nature Reviews. Drug Discovery, 10*(9), 712. https://doi.org/10.1038/nrd3439-c1

Rijcke, S. d., Wouters, P. F., Rushforth, A. D., Franssen, T. P., & Hammarfelt, B. (2015). Evaluation practices and effects of indicator use—A literature review. *Research Evaluation, 25*(2), 161–169. https://doi.org/10.1093/reseval/rvv038

Rip, A. (1994). The republic of science in the 1990s. *Higher Education, 28*(1), 3–23. https://doi.org/10.1007/BF01383569

Rip, A., & van der Meulen, B. J. R. (1996). The post-modern research system. *Science and Public Policy, 23*(6), 343–352. https://doi.org/10.1093/spp/23.6.343

Snow, C. P. (1993). *The two cultures.* Cambridge Unversity Press.

Stephan, P. E. (1996). The economics of science. *Journal of Economic Literature, 34*(3), 1199–1235. Retrieved from www.jstor.org/stable/2729500

Stokes, D. E. (1997). *Pasteur's quadrant: Basic science and technological innovation.* Brookings Institution Press.

Toulmin, S. (1972). *Human Understanding.* Clarendon Press.

Van der Meulen, B. J. R. (1997). The use of S&T indicators in science policy: Dutch experiences and theoretical perspectives from policy analysis. *Scientometrics, 38*(1), 87–101. https://doi.org/10.1007/BF02461125

Wei, X., Ghosh, S. K., Taylor, M. E., Johnson, V. A., Emini, E. A., Deutsch, P., et al. (1995). Viral dynamics in human immunodeficiency virus type 1 infection. *Nature, 373*(6510), 117–122. https://doi.org/10.1038/373117a0

Whitley, R. (2000). *The intellectual and social Organization of the Sciences* (Vol. 11). Oxford University Press.

Whitley, R., & Gläser, J. (2007). *The changing governance of the sciences: The advent of research evaluation systems.* Springer.

Wilsdon, J. (2016). *The metric tide: The independent review of the role of metrics in research assessment & management.* SAGE.

Winch, P. (1958). *The idea of a social science and its relation to philosophy.* Routledge and Kegan Paul Humanities Press.

Wolthers, K. C., Bea, G., Wisman, A., Otto, S. A., de Roda Husman, A. M., Schaft, N., … Miedema, F. (1996). T cell telomere length in HIV-1 infection: No evidence for increased CD4+ T cell turnover. *Science, 274*(5292), 1543–1547. https://doi.org/10.1126/science.274.5292.1543

Wouters, P. F. (1999). *The citation culture.* University of Amsterdam.

Wouters, P. (2014). The Citation from Culture to infrastructure. In Cronin, B., & Sugimoto, C. R. (Eds.), *Beyond bibliometrics. Harnessing multidimensional indicators of scholarly impact* (pp. 1 online resource (viii, 466 pages)). MIT.

Wouters, P. (2017). Eugene Garfield (1925–2017). *Nature, 543*(7646), 492–492. https://doi.org/10.1038/543492a

Wouters, P. F., Thellwall, M., Kousha, K., Waltman, L., De Rijcke, S., Rushforth, A.D., Franssen, T. (2015). *The metric tide. Supplementary report 1 to the independent review of the role of metrics in research assessment and management.* Commissioned by the Higher Education Funding Council for England (HEFCE). Retrieved from https://re.ukri.org/documents/hefce-documents/metric-tide-lit-review-1/

Ziman, J. M. (1978). *Reliable knowledge: An exploration of the grounds for belief in science.* Cambridge University Press.

Chapter 4
Science for, in and with Society: Pragmatism by Default

Abstract To rethink the relation between science and society and its current problems authoritative scholars in the US and Europe, but also around the globe, have since 1980 implicitly and increasingly explicitly gone back to the ideas of American pragmatism. Pragmatism as conceived by its founders Peirce, James and Dewey is known for its distinct philosophy/sociology of science and political theory. They argued that philosophy should not focus on theoretical esoteric problems with hair-splitting abstract debates of no interest to scientists because unrelated to their practice and problems in the real world. In a realistic philosophy of science, they did not accept foundationalism, dismissed the myth of given eternal principles, the unique 'scientific method', absolute truths or let alone a unifying theory. They saw science as a plural, thoroughly social activity that has to be directed to real world problems and subsequent interventions and action. 'Truth' in their sense was related to the potential and possible impact of the proposition when turned in to action. Knowledge claims were regarded per definition a product of the community of inquirers, fallible and through continuous testing in action were to be improved. Until 1950, this was the most influential intellectual movement in the USA, but with very little impact in Europe. Because of the dominance of the analytic positivistic approach to the philosophy of science, after 1950 it lost it standing. After the demise of analytical philosophy, in the 1980s of the previous century, there was a resurgence of pragmatism led by several so-called new or neo-pragmatists. Influential philosophers like Hillary Putnam and Philip Kitcher coming from the tradition of analytic philosophy have written about their gradual conversion to pragmatism, for which in the early days they were frowned upon by their esteemed colleagues. This new pragmatist movement gained traction first in the US, in particular through works of Bernstein, Toulmin, Rorty, Putnam and Hacking, but also gained influence in Europe, early on though the works of Apel, Habermas and later Latour.

In the previous chapter I discussed the problems and distortions of the practice of scientific inquiry and of the organization of academia. These problems do not only affect <u>how</u> we do research, but also <u>which</u> research is being done or not done. The latter is what philosophers and sociologists designate the growth of knowledge, or to use Longino's phrase 'the fate of knowledge'. The latter reminds us that knowledge claims can be reconsidered and refuted, but we also can think of knowledge that never was. It was not produced because the required inquiry it was decided, or even never consciously considered nor decided, not to be pursued. This is, as discussed in Chap. 3, not the classical problem of the 'invisible hand', but directly reflects the politics of science and research in academia influenced by idiosyncratic or otherwise motivated scientists, public and private funders and government agencies. This process is operated by very 'visible hands' belonging to a large number of individuals who are serving on boards and advisory or grant committees like NIH study sections of the many institutions and organizations of the science system, at the institutional, national and international level. These organizations still dominantly use frontstage narratives that largely originate from two major sources one from the inside and the other from the outside. From the inside it was the Legend with all the preconceived ideas and its consequences, discussed in the previous chapters, and from the outside it was the capitalist ideas of economic power and profit and of technological control in the modern knowledge society. These two ideologies have in strong synergy since the 1980s shaped scientific inquiry with serious consequences for the growth of knowledge and thus for society at large and at the personal level for the lives we live. This is experienced daily and is being increasingly recognised by virtually all researchers in the international scientific community, which in the past ten years has led to a global discussion of how science is broken and how to improve or, if possible, fix it.

Before going to discuss the prospects, opportunities, pitfalls and dangers for change of the aims and institutional organization of scientific inquiry, I will take a step back and reflect on the consequences of the conclusion of the previous chapter that the academy and the practice of research are in need of serious change. To successfully make this change happen, a series of essential changes have to be made, with the required precautions taken, that will gradually promote and enable the required transition in the coming years. For this transition we need to understand (as Hacking would say, 'take a good look at') the modern practice of science, how it is done in the daily life of researchers, but also how researchers off the record talk amongst each other about what they do. It is as important to understand how in general the community of researchers in fosters a particular image of science especially when talking to lay audiences and when scientists or science administrators appear in the media. It has become clear in the previous chapters that the popular image is a Legend and does not at all match with what the practice of science is and how research is (and was) done. That classical myth, although obsolete and untenable, is still dominant, and more important, inhibitory to the required change to make science and research fit for the future. In this chapter I will argue for a powerful alternative theory and vision. This new narrative needs to provide a modern, more social and humanistic image which must have a firm basis in modern thinking in philosophy, history and sociology of science. This modern image must thus not be a myth like the Legend but must be being recognized and practiced by active researchers and be an all-encompassing empirical account and theory about the many different styles and practices of scientific and scholarly research, in the past and at present.

Several writers about science who have discussed the practice of science have described new modes of science as it has developed since the 1980s or should develop, using labels like 'industrialized' (Ravetz, 1971) and 'post-normal' (Ravetz, 2011), 'post-academic' (Ziman, 1994, 2000), 'Mode-2' (Gibbons et al., 1994).

Ravetz (1971), Ziman (2000) and Nowotny et al. (2001) do elaborate on a general practical theory and philosophy of science which explicitly refers to recent developments in the philosophy and sociology of science and more broadly in STS research after 1980. As discussed above, until the early 1990s, the truly multidisciplinary Science and Technology Studies, which is what these writers practiced, had still to come of age as a respectable academic discipline in its own right. Mainstream professionals guarded the tribal fences between philosophy, sociology, anthropology, psychology, economics and history of science. This has not promoted our understanding of the interrelationships of society and science in history and modern times. Toulmin as we saw complained about it in the late 1950s. Even Bruno Latour who notoriously and successfully has been crossing these borders, which he argues is common in anthropological studies of tribal life in the Amazon or New Guinea, complained in his *'We have never been modern''* about these dualistic seams between nature, culture and the sciences. *'We pass from a limited problem - why do the (sociotechnological) networks remain elusive? Why are science studies ignored?- to a broader and more classical problem: what does it mean to be modern?- When we dig beneath the surface of our elders; surprised at the networks that- as we see it- weave our world, we discover the anthropological roots of the lack of understanding. Fortunately, we are being assisted by some major events that are burying the old critical mole in its own burrows. If the modern world in its turn is becoming susceptible to anthropological treatment, this is because something has happened to it. ....*

we have known that it took a cataclysm like the Great War for intellectual culture to change it habits slightly and open its doors to the upstarts who had been pale before (Quotes are from section 1.3) (Latour, 1993).

Although Latour continues this observation with the fall of the Berlin Wall in 1989, another cataclysmic global shock, for me his lines take us back to Chap. 1, and John Dewey who at the beginning of the twentieth century in fact before and after the Great War of 1914–1918, argued for another science and philosophy, that both are socially and culturally more inclusive and reflexive than the (natural) sciences of his days. It is pragmatism that recently many became to believe provides the best approach for our understanding of and contributing to the complex of society and science. The pragmatist theory of scientific inquiry, developed predominantly between 1870 and 1940 by the early American pragmatists, Peirce, Dewey and James, was rejuvenated and modernized by a group of high-profile 'new pragmatists' (Misak, 2007) in the second half of the last century. In this chapter, I will briefly discuss the essential features of pragmatism and argue that it provides the default theory and concepts of the aims and practice of science since it is open, non-dogmatic and pluralistic, inclusive and contextual, lives up to our present state of hyper-modernism and acceleration with fluidity of place and time. For the researchers it does not provide a mythical 'scientific method/idealistic, positivist Cartesian certainty based on formal rules and foundations, however, it does provide rich guidance and understanding of the objectivity of the reasoning, and functioning of the processes and practices of the communities of inquiry in the sciences and humanities and how they may change over time. Pragmatism is clear about the intersubjective procedure of evaluation of our accepted scientific beliefs, importantly in applying and testing them in actions and interventions and is honest about the intrinsic fallibility of our beliefs. Pragmatism is fallibilistic but is in essence against scepticism. Scepticism may be a fine attitude for the study room, academic debates and for papers, but it loses force in the outside world.

Pragmatism in principal sees scientific research as a means to an end. The ultimate aim is to address and alleviate problems and issues that prevent people from living the good life. Therefore, science must constantly engage with the publics and their problems and science is thus seen a key component of the aspiration of the true idea of democracy, not naive but realizing all its issues (Dewey & Rogers, 2016). This all-encompassing concept of theory and practice can provide guidance for shaping the organization of modern science and inquiry, of aims and ownership and the common good, participation, processes, ideal deliberations and agenda setting, inclusive evaluation criteria - incorporating facts, values and goals, action, interventions and implementation- and social reflexivity of all these steps which is needs [inside] because it hits us from [outside] from an ever more rapidly changing hyper-reflexive modern society (Beck et al., 1994; Nowotny et al., 2001). Outside and inside in the previous sentence were put in brackets because the classically defined boundaries between science and society, the experts and lay publics, are and have always been permeable, which was experienced with a negative connotation as leaky by those who had held on to the dualities of the scientific method of the Legend.

4.1 Pragmatism by Default

Given these considerations about science and society, and the demise of analytical philosophy, there are two main reasons why it is believed that pragmatism with the diverse new pragmatist interpretations of recent times is the best idea of science and philosophy of science. First, it provides insight and understanding which matches the practice of science since it starts from a realistic historical and sociological understanding of the social practice of science. Second, for philosophers of science who are active in the post-empiricist positivist era, pragmatism appears to be an acceptable and fruitful philosophical proposition that is not impeded by esoteric problems as empiricism and positivism both are.

For the philosophers who started their training and academic careers before the 1960s or even 1970s and whose philosophical thinking until late in life has been dominated by positivism or various kinds of theories of empiricism and realism, this 'pragmatic turn' has not been easy. All of them in their articles and books literally describe it as a process of conversion, a paradigm shift which was frowned upon or ridiculed by their peers and colleagues.

Most writers about pragmatism describe how pragmatism in the USA was dominant until the 1930s but was rapidly overtaken by the analytical tradition (Diggins, 1994; Misak, 2013). With the rise of the analytic positivistic tradition -which the members of the Vienna Circle, after fleeing Europe in the 1930s have spread across the USA- pragmatism rapidly lost its influence. As we have seen (in Chap. 2), mathematics and the natural sciences as dominant models for science and the Cartesian system of dualisms shaped the analytical and linguistic turn in philosophy of science. In the eyes of the diehard philosophers of those days, compared to the rational and formal approach of logical positivism and empiricism, pragmatism had little of epistemology and of a formal philosophical system to offer. Peirce's philosophy was nearest to such a system with his analysis of the three methods of inference: induction, deduction and abduction. Because of this, Popper and the Popperians and some philosophers who came through the analytical tradition, like Nagel, Putnam, Hacking and later Misak, had strong affinity for Peirce. James and Dewey did not bother with that formal philosophy and explained their thoughts and argued and reasoned in plain language. The new-pragmatists, like Rorty, Bernstein and Kitcher where more engaged with James and Dewey's broader view about the social and political, 'science in democracy' as Kitcher called it. Make no mistake, this writing style, reasoning and argumentation devoid of the esoteric 'analytical-logical-formal' however, is misleading regarding the depth of thought and insight proffered, as Putnam said about Dewey and Hacking said about Peirce and James.

4.2 Why Bother?

I believe that for practising scientists, both natural scientist and scholars from SSH, who have a certain degree of proper self-understanding of their methods, the meaning of its intersubjectivity, the limitations of its claims and the social aspects of their practice of inquiry, pragmatism may well be considered a most realistic image and theory of their daily work. Moreover, even those who have not reflected a lot, or young professionals who not yet thought a lot about these issues, which we know is not unusual at least in the biomedical and natural sciences, pragmatism may come on to them as quite naturalistic descriptive. What then does pragmatism have to offer to them? Given what I discussed in Chap. 2, in the confined space of the practice of inquiry, studying and researching, doing experiments and interventions -in the library, the lab, the clinic or in societal practices- not too much. At that level, scientists, do adhere to validated and accepted methods, logics and procedures of their respective disciplines, but do not bother on a daily basis with the higher levels of philosophical assumptions. So why should they, or we, now bother about pragmatism? They, and we, should very much care about pragmatism. We have seen in Chap. 3 that only one level up, where the mundane matters of management like strategy, policy and governance are discussed, the assumptions of the Legend still reign. This is most visible as soon as we have to consider issues of quality, excellence, acceptability, impact, and evaluation. Then assumptions of the Legend immediately become visible and are at the table in the deliberations which sets scene and tone and in part cause the distortions discussed in Chap. 3. It is at this level that pragmatism will provide realistic guidance for these deliberations and agenda setting, inclusive evaluation criteria - incorporating facts, values and goals, action, interventions and implementation- and social reflexivity of all these steps. At an even higher level, it likewise can be instrumental for shaping the mission and strategy of the organization and government of science at institutional and national level, regarding its higher purpose, aims and ownership and relation to the wider public. At this level, pragmatism because of its realistic, modern, open and democratic view of science, allows for a better narrative with responsibility in how we communicate about science and research to and importantly engage with the various public representatives and public debates and in the media. What forces were working against the pragmatic turn?

Reading the vast body, or even the top 10% of the literature of the past 30 years on pragmatism and the pragmatic turn is impossible and I believe not required for the argument to be made in this book. There is, paradoxically already a lot of esoteric writing about these philosophers whose thesis it was that philosophy should not deteriorate into esoteric writing that does not bother anybody in the real world anymore. The secondary literature on the classical philosophers, Russel, Popper, Kuhn and Wittgenstein, the famous Vienna Circle and the Frankfurt School is also vast, and many have read not so much the original texts but overviews in the books about philosophy of science and modern science. Until very recently, textbooks of the philosophy of science, even philosophy of the humanities, social science and even sociology rarely mention or discuss the work of the early pragmatists, some do refer to Rorty's progressive interpretations of James and Dewey. When I recently confronted some well-known Dutch authors of these textbooks who I knew clearly do sympathize with pragmatism with this omission, they shrug their shoulders. They reply with the words 'I thought it was not yet philosophically developed enough' or that 'it is not yet suitable for introductory texts books'. Instead, we offer our students mainly still the myth of The Legend vintage 1950s, with sometimes a small side dish of Kuhn vintage 1962 and a glims of the early works of Latour vintage 1979 or 1983 with the explicit warning 'watch out it's spicy'.

Barker and Kitcher, however in their very nice textbook *Philosophy of Science* (2013), where the demise of logical positivism is spelled out, if not celebrated, discuss how we are now able to come to a realistic image of the pluriform practices and can be frank about the limitations of the sciences. Even there, no reference to an alternative realistic narrative of pragmatism is to be found (Barker & Kitcher, 2013). This is of interest given the life-long struggle of Kitcher with his conversion described in Chap. 2. In his *'Preludes to Pragmatism'* and *'The Ethical Project'* written in the same years as his 2011, he takes Dewey's pragmatism as the leading philosophy to think about modern science and ethics in democracy (Kitcher, 2011, 2012). Is it really the case that main-stream philosophers, sociologists and science and technology scholars writing about science consciously kept a safe distance to pragmatism because intellectually and emotionally the gap between the Legend and pragmatism was too big for them? Yes, and Kitcher is most frank about it on the very first pages of *Preludes*: '*Classical pragmatism is, I believe, not only America's most important contribution to philosophy, but also one of the most significant developments in the history of the subject...'Twenty years ago, I would not have made that judgement. Like most of my contemporaries in philosophy departments in the Anglophone world, I would have seen the three canonical pragmatists -Peirce, James and Dewey- as well-intentioned but benighted, labouring with crude tools to develop ideas that were far more rigorously and exactly shaped by the immigrants from Central Europe whose work generated what is (unfortunately) known as "analytic" philosophy.'*pxi (Kitcher, 2012).

Because of this it has not resulted in a reform and its influence faded apart from a few philosophers who have kept the debate about it going. Is the pragmatic turn difficult, for them and most of us, because pragmatism does not offer a new myth or fresh ideology for the twenty-first century which provides a sense of certainty, an uncontested foundation, a legitimation with which we can assure ourselves and the public about the authority of science? Given Dewey's severe criticism about this *quest for certainty* and the history of the demise of the Legend the deceptively common-sense philosophy of pragmatism clearly seems to contribute to the uncomfortable relation the philosophers and interested scientists have with pragmatism. In addition, we have seen that the Legend has had enormous impact on the politics of science in relation to society, as frontstage narrative, but that this narrative paradoxically is even in use within science, backstage (!) and there has distorted the general view of the sciences and the humanities. At both these levels of the scientific community the pragmatic turn thus will surely bring gains to many, but losses to others, the former academic elites that lose reputation, access to control and power and its many associated advantages. This institutional feeling of loss and uncertainty also may hold for the philosophers who did not want to be affiliated with non-mainstream philosophy and their proponents. Is it so that only after a successful mainstream professional career in academia, with independence of one's peers, there is finally room and opportunity to engage with the non-mythical mundane pragmatism and does one have the guts to be frank about the Legend?

I will restrict myself here to a concise overview of the main concepts of pragmatism and discuss a bit more in detail the more recent works of the new pragmatists as far as it relates to the philosophical principles and ideas of inquiry.

Richard Bernstein, whose perspective is from the humanities and social sciences, and his experience in the US liberal arts college system, has written with great authority from the broader pragmatist perspective (Bernstein, 1983, 2010). His *Overview* (p1–49, (Bernstein, 1983)) is quite technical, but provides a comprehensive history of the concept of rationality in modern philosophy which makes the strong case that pragmatism is the default (in my words). His discussion of the work of Habermas and the early influence of Peirce on Habermas (Habermas, 1970, 1971) will be revisited in Chap. 5. Hacking's *Representing and Intervening* (Hacking, 1983), I have quoted already, is very concisely discussing the problems of positivism, especially in the three last pages were the legacy of Peirce is brought in, and in the bit more than 5 pages on what pragmatism has to offer. Hillary Putnam's *Pragmatism* (Putnam, 1995) especially the less technical chapters on

William James and the in total 18 pages of Chap. 2 on *Pragmatism and the Contemporary Debates* are very good reads.

The Metaphysical Club, by Louis Menand a professor of English, is a prize-winning highly praised, more literary intellectual history of pragmatism (Menand, 2001). It discusses what prompted these thinkers to work out this unique truly American philosophy between 1870 and 1940. He describes quite colourfully, how they differed in the range of issues they wrote about, in their style and temperament and political engagement. We also get a view of the very different sometimes deeply troubled personal lives they have had, which especially relates to Peirce. Reading this book makes you realize how different the world and the philosophical, religious and political issues were only one hundred years ago. At the same time. it becomes clear how modern and humanistic the pragmatists were regarding their ideas about scientific inquiry, their critique of the Cartesian and positivistic philosophies, the relation with society and the publics, the methods and social structures. It becomes clear that they reflected on inquiry not only from the point of view of *episteme* (theoretical knowledge), but also *techne* (technological application and action) and *phronesis* (practical wisdom and reason) (Bernstein, 1983). Dewey later started a real pragmatist movement which took the thinking and philosophy to many other fields of humanities and social sciences, most of all educational theory, ethics, and political theory on for instance the workings of democracy in the Chicago Laboratory School. Menand provides a fine accessible summary of pragmatism in non-technical language in chapter 13 (p351–375), which starts as follows:

> *Pragmatism is an account of the way people think, the way they come up with ideas, form beliefs, and reach decisions'...there is no noncircular set of criteria for knowing whether a particular belief is true. No appeal to some standard outside the process of coming to the belief itself.*

He cites James who had the most expressive style of writing and has been instrumental in promoting the work of Peirce and Dewey in the USA: *'Truth happens to an idea. It becomes true, is made true by events. Its verity is in fact an event, a process, the process namely of verifying it.' 'Beliefs, in short, are really rules for action, and the whole function of thinking is but one step in the production of habits of action.'* He cites James' most discussed and debated statement, which takes the philosophy of Peirce in the eyes of Peirce much too far, but has much inspired Rorty sixty years later years later: *'..the true is the name of whatever proves itself to be good in the way of belief'*. This could in our days well have been a tweet.

Charles Sanders Peirce (1839–1914) was the real founder of pragmatism in the eyes of most philosophers of science. What he thought and wrote at the age of 29 in 1870 is most impressively reflecting unbelievable intelligent, independent broad and original scholarship. Reading about him, his temperaments, his personal problems, the hardships that befell him, and how that has also affected his professional career makes you feel sad. Peirce was trained as a natural scientist with laboratory experience and made major contributions to mathematics and formal logics and is regarded as one of the most brilliant American philosophers (Nagel, 1940). He showed us the way out of Cartesian dualisms, the dichotomy of fact and value, the problem of representation by theory of reality and the problem of foundations and 'truth'. The impact of his work outside the US was recognized in England by Frank Ramsey in the 1920s. Ramsey discussed Peirce' philosophy with Russel and Moore and in several sessions with Ludwig Wittgenstein in Vienna. (Misak, 2013; Putnam, 1995) Peirce also influenced Popper who agreed with 'his critique of the search for epistemological origins that has dominated so much of modern philosophy' (Bernstein, 1983, 2010). Bernstein emphasizes that Peirce, next to more method-ological ideas, has strongly proposed the concept of <u>the community of inquirers</u> and *'his relentless criticism of the subjectivism that lies at the heart of so much modern epistemology'* and connects to modern major influential thinkers, in the next lines: *'...he develops an intersubjective (social) understanding of inquiry, knowing, com-munication, and logic. Jürgen Habermas has argued that at the turn of the twentieth century there was a major paradigm shift from a 'philosophy of subjectivity' or a 'philosophy of consciousness' to an intersubjectivity (social) <u>communicative model of human action</u> and rationality. One of the primary sources of this shift is evident in Peirce' early papers. The above passage also anticipates the centrality of the community of inquirers in Peirce's pragmatism. ..To say that inquiry is self-correcting is to say that a critical community of inquirers has the intellectual resources for self-correction.'*

> *It is only in and through subjecting our prejucies, hypotheses, and guesses to public criti-cism by a relevant community of inquirers that we can hope to escape from our limited perspectives, test our beliefs and bring about the growth of knowledge* (p35/36) (Bernstein, 1983).

We have seen in the previous chapter that this is an ideal of integrity, a major critical aspiration that the community has to effectively perform at all levels of inquiry. Peirce and especially Dewey have been criticized as being naïve in their views of communication and interactions in the process of inquiry and, in Dewey's later works, engaging publics from outside academia. Popper (Popper, 1981) who also in the same vein emphasized the continuous process of criticism in science, also warned against distortions of the discourse by internal and external interests. As the founder, or one of the founders of pragmatism Peirce is favoured and admired especially by philosophers who came from the analytic tradition. Above I cited James's popular version, we would now say 'tweet' of the pragmatic maxim, but Peirce as originator of the maxim was much more subtle on this. Misak, but also others have tried to correct the popular view that was instigated by James. Misak (p29) writes that his notorious statement is to be understood as follows*: 'Consider what effects, which might conceivable have practical bearings, we conceive the object of our conception to have. Then, our conception of these is the whole of our conception of the object'. ... 'we must look to the upshot of our concepts in order to rightly comprehend them'.* And Misak's favourite: *'we must not begin by talking of pure ideas, – vagabond thoughts that tramps the public roads without any human habitation, – but must begin with men and their conversation'* Misak p31 (Misak, 2013).

He rejected given, timeless principles, stating that 'there is no cognition "not determined by a previous cognition" or "something outside of consciousness" (p39). *"he thought that 'truth was a matter for the community of inquirers' not for the individual inquirer'. Science and inquiry and rationality are matters of getting our beliefs in line with experience, evidence and reason in an ongoing community project. In our efforts to understand reality "each of us is an insurance company'* (p37). This process in practice does never stop. Peirce is categorical to state that this is the case since all our beliefs are imperfect and are subject to continuous testing. There are degrees of acceptance and of trust in a belief of course. It is, he proposes, by this process of *Fixation of Belief* we gradually improve and finally come to a set of converging true beliefs. But when is finally? This is problematic but not really: like Popper, Peirce proposed a metaphor: *'its reasoning should not form a chain which is never stronger than its weakest link, but a cable whose fibres may ever be so slender provided they are sufficiently numerous and intimately connected* (Collected Papers 5.265). He proposed no unique method, but deduction, induction and abduction, also designated inference to the best explanation. This reminds us in many respects of Poppers falsificationism of conjectures and refutations. It does sound familiar to active scientists in the natural and social sciences with respect to the hypo-deductive method starting with an idea, or hypothesis to be tested and proven, but falsification and refutation is not really the main goal in daily in practice.

John Dewey, a student of Peirce, was very much inspired by the work of Peirce but his view of philosophy and scientific inquiry was much broader. He was concerned with the role of science in the broad scheme of the problems of society and its diverse publics and of democracy. He wrote extensively about the relation between science, the conduct of inquiry and the problems of these publics. This philosophy which naturally flows over in political theory is the pragmatism I will discuss in Chap. 5 where involvement and engagement of the publics with scientific inquiry will be discussed in terms of the present societal challenges in our modern times. Dewey had a background in educational theory and pedagogy, child upbringing and development. In his thinking education was a major factor in building civic communities that could allow for public to participate in deliberation about inquiry and action. Education in his mind was life itself. For him inquiry must be prompted by a concrete situation of doubt or a problem and thus foremost had the obligation to contribute to mitigation or solving issues that hindered people from leading the good life. This was the short-term aim of science and he did not bother with the Peircean epistemological problems how in ongoing inquiry in the long run 'truth' comes about. Dewey was a true public intellectual who connected in a natural way inquiry with social action in which he himself engaged forcefully in political actions. He had high visibility in American public life, politics and its debates for instance at the times of McCarthyism.

Bernstein elaborates on Dewey's vision of radical democracy which will be revisited in Chap. 5 (Bernstein, 1983). Dewey wrote widely and a lot. His contribution to the philosophy of science which is most relevant here has been summarized by Hacking where he divides pragmatism in two *'Peirce and Putnam on the one hand and James, Dewey and Rorty on the other. ..It is interesting, for Peirce and*

*Putnam both to define the real and to know what, within our scheme of things, will
pan out as real. This is not of much interest to the other sort of pragmatism. How we
live and talk is what matters, in those quarters. There is not only no external truth,
but there are no external or even evolving canons of rationality. Rorty regards all
our life as a matter of conversation'. Dewey rightly despises the spectator theory of
knowledge...the right track in Dewey is the attempt to destroy the conception of
knowledge and reality as a matter of thought and* **representation***. He should have
returned the minds of philosophers to experimental science... in his opinion things
we make (including all tools, including language as a tool) are instruments that*
intervene *when we turn our experiences into thoughts and deeds that serve our
purposes..' (p 62/63) (Hacking, 1983).*

Putnam, whom I introduced in Chap. 2, has made an intellectual journey from
analytical philosophy to pragmatism, and even after 1981 apparently became more
influenced by the works of James and Dewey than Hacking in 1983 had anticipated.
In the collection of papers published with the telling title *Words and Life (*Putnam,
1995; Putnam & Conant, 1994*)* there is deep admiration for Dewey's philosophy of
inquiry as shown in *Pragmatism* that same year. '*Perhaps the most detailed case for
the view just defended, the view that all inquiry, including in pure science itself
presupposes values, is made by Dewey in his Logic* (Dewey, 1939), *here I want only
to discuss one aspect of Dewey's view, the insistence on a very substantial overlap
between our cognitive values and our ethical moral values. I have already examined
the claim that there is a fundamental ontological difference between cognitive or
'scientific' values, and found that the reasons offered for believing that claim fail.'*

Comparing Carnap's (positivistic) view with Dewey's: '*For Dewey, inquiry is
cooperative human interaction with an environment; and both aspects, the interactive intervention, the active manipulation of the environment, and the cooperation
with other human beings, are critical. For the positivists...the most primitive form
of scientific inquiry, and the form that they studied first when they constructed their
(otherwise very different) theories of induction, was by simply enumerating. The
model is always a single scientist...For Dewey the model is a group of inquirers trying to produce the good ideas and trying to test them to see which ones have value'.*

Putnam then states: *...cooperation must be of a certain kind in order to be effective,. It must, for example, obey the principles of "discourse ethics" [here he cites
Habermas]...When relations among scientists become relations of hierarchy and
dependence, or when scientists instrumentalize other scientists, again the scientific
enterprise suffers.'* Dewey was as Putnam states, *not naïve and was aware that there
are power plays in the history of science as in the history of every human institution,
'"but he still holds that it makes sense to have a normative notion of science....Both
for its full development and for its full application to human problems, science
requires the* <u>democratization of inquiry</u>.

"*Dewey opposes the of the philosophers 'habit of dichotomization of inquiry.' in
particular he opposed both the dichotomy "pure science/applied science" and the
dichotomy 'instrumental value/terminal value". Pure science and applied science
are interdependent and interprenetrating activities, Dewey argues. ...Science helps
us to achieve many goals other than the attainment of knowledge for its own sake,*

and when we allow inquiry to be democratized simply because doing so helps us achieve those practical goals, we are engaged in goal-oriented activity. ..we are not- nor ever were-interested in knowledge only for its practical benefits; curiosity is coeval with [= as old as] *the species itself, and pure knowledge is always, to some extent, and in some areas, a terminal value even for the least curious among us* (p172, 173) (Putnam & Conant, 1994).

I have in the previous chapter demonstrated, using Bourdieu's theory of 'the field'(Bourdieu, 1975), how the internal politics and power games of science have in the past 40 years developed into a system where the discourse ethics due to, among others these dichotomies of the Legend and other related interests is heavily plagued if not seriously distorted. I will discuss in the next chapter how I think the community of inquirers can be improved and organized based on these insights.

4.3   New Pragmatists

Philip Kitcher is widely considered to be one of the leading figures of contemporary philosophy of science. I have in Chap. 2 referred to his profound intellectual struggles to release or even to liberate himself of the analytical tradition and the myths of the Legend. In his *Science, Truth and Democracy (*Kitcher, 2001*)*, he takes his critique of the Legend quite some steps further than in *The Advancement of Science* published only eight years before.(Kitcher, 1993) His phrasing is cautious, given the then still raging 'science wars' about foundations, objectivity and scientific authority, in order *'to articulate a picture of the aims and accomplishments of the sciences so that moral and social questions can be brought into clearer focus'* (p xii). He discusses in the first six short chapters the claims and problems of the Legend related to objectivity, theory choice and how next to cognitive values, social and ethical values play a role gradually introducing the context of inquiry: aims, theoretical and practical interests and social, moral, political and religious values. It feels as if he wants to take the believers of the Legend by the hand and lead them through the desert (of the demise of the Legend) to the other side where between the extremes fertile soil await scientific inquirers, no matter if they are from the 'hard' or 'soft' sciences. In this book, Kitcher does not explicitly tell the reader that this fertile soil is to be found in the land of pragmatism. Rorty, who wrote like Dupré some nice lines on the back of the paperback edition of 2003, put it like this *'Kitcher navigates very skilfully between the extremes of positivistic science-worship and Foucauldian distrust of the regimes of truth'.* Kitcher, reflects on representation and interventions (p52):*'Representations are constructed, but do not construct the world'.* But...*' the impact of categories [claims and theories] on reality 'by way of human intervention is more evident in the biological sciences than in the physical sciences and most striking in those areas of inquiry in which we study ourselves.' 'Categories are consequential. Accordingly, there is important work ....to do be done in reconstructing the ways* *in which our most influential divisions* (ideas about reality) *were constructed and how they have left their mark on the world we inherit'*.

The history of the 'hard' sciences provides excellent examples for this, but Kitcher mentions scientifically derived labels such as 'insanity', 'race' 'homosexuality', we now would add 'inequality', 'health' or from a reflexive viewpoint 'absolute truth obtained by pure scientific inquiry'. Kitcher refers to the theoretical and social critique of Michel Foucault, a stranger in the land of the Legend, but thriving on the fertile soil of pragmatism. In the next step Kitcher moves to the idea that theories are to be regarded as maps, a powerful metaphor explicitly put forward before by Wittgenstein, Toulmin and Ziman. Maps are not to be taken to literary reflect the world, but always be a substitute that, when accurate, of great value to assist us in navigating and intervening in the world. To take actions, humans trust accepted beliefs that have been shown and proven to work. Users of a specific map, can improve the map (or have the map improved) based on problems they experienced when they used it, applying new knowledge and technology. There are at the same time many maps possible of a given territory dependent on the changing interests of its users and with new knowledge other maps will be produced. These are different, but not per definition better maps. The key question for us now is: Who does, and how do we define what is good? With the metaphor of the map Kitcher arrives at the question of *'the goals of inquiry, a specification of what constitutes significant science that will apply across all historical contexts and, independent of the evolving interests of human beings'* p62). Indeed using examples from biology, he concludes: *'Like maps, scientific theories...reflect the concern of the age. There is no ideal atlas, no compendium of laws, or "objective explanation" at which inquiry aims'*. In an interesting intermezzo, the issue of value neutrality, autonomy and academic freedom, classical flaws of the Legend are addressed. These nearly six pages are thus very relevant for our discussion of the myth and how an alternative more realistic theory of scientific inquiry may help out. Kitcher approaches the problem via what he calls 'the myth of purity', the 'pure versus applied dichotomy' rejected on conceptual grounds by scholars before (see Chap. 1).

Kitcher, agrees with Dewey that science for the sake of science, to add to the body of knowledge can be significant on pure cognitive grounds. *'The aim of science (pure science, basic research) is to find truth; the aim of technology (applied science) is to solve practical problems.'* But it is not that simple: *'the aim of science is to discover significant truths'* (p87). It is recognized that there is always a chance of practical use somehow, but this is not the interest of the investigator who says to pursue curiosity for curiosity sake. Indeed, we know that, although these investigators virtually all proudly state that they do pure fundamental science, as soon as they are interviewed about their work because of a Nobel prize, a breakthrough paper or a major personal grant that they have won, they start to explain how their work may lead to a new method of treatment, medicine, help solve problems of green energy, etc. There are, says Kitcher always motives for the 'pure' scientist in the background or actually, as we saw in Chap. 3, in the foreground, as reputation, fame, career options, access to funding because of discoveries to be made. That is not different for pure or applied science and not even for technical sciences. The distinction between pure and applied research is blurry, complex and *'in extreme cases researchers can quite legitimately declare their intentions to be thoroughly epistemic. However, when only little curiosity is needed to see that the current [knowledge] has been shaped by dubious ventures from the past, or when the propensity of others to engage in morally consequential applications ought be obvious, the researcher who proclaims solely epistemic intent is guilty of self-deception (at the very least). Pure researchers, then are not simply whose intentions are entirely to promote epistemic significance but whose lack of interest in the practical can be justified'*(p89) Why has the distinction seemed so important? *'It seems to be to limit the scope of moral, social, and political appraisal...to which the practice of science is accountable....but only in the context of applied science, or of technology. The myth of purity proposes that there is a distinction that fulfils these purposes.'*(p89-90) Kitcher obviously rejects this myth for this specific reason of neutrality and evasion of responsibility. We have seen in the previous chapters that the myth of purity also implicitly and explicitly confers the message to and from academia that pure science is morally and ethically pure and therefore the 'high church', whereas applied science and technology are stained with non-scientific bias and interests hence are 'low church'. This goes back to Greek philosophy and has survived until it was incorporated in the Legend but is firmly rejected by pragmatism. It are the intentions, the value and impact, the actions it makes possible, not the practices and methods of inquiry that count.

In the following chapters, he takes the final step, beyond *'the traditional philosophy of science .. that provided a very narrow normative perspective science.'*(p111) The next problem is how to organize *well-ordered science* in the larger community of inquiry in interaction with policy making and the publics knowing that for significant science it must relate to contexts where the problems are. This is a central theme encompassing all sciences alike pure, applied and technology. Kitcher agrees with the pragmatists that it is the obligation and responsibility of scientists to strive for well-ordered science. The interests of the less powerful publics are to be cared for, taking into account as Kitcher states, the problem of vulgar democracy and tyranny of the ignorant which is a nightmare scenario for scientists who believe any interaction with representatives of lay publics threatens basic science. On the other hand, scientific inquiry needs to be protected against the interests of the powerful private parties who have advantages in funding and protected from unwanted political influences. A problem that has grown bigger and bigger since 1945.

In his *Science in a Democratic Society* published in 2011, Kitcher again takes on the problem of the ideal of well-ordered science in a well-ordered society, obviously much inspired by Dewey. In this book, Kitcher apparently deliberately abstains from explicitly presenting pragmatism as an alternative, or as I say, the default for the obsolete flawed views of the Legend. Still, Kitcher states that the problems of science, in his opinion, relates to the classical theoretical picture of scientific inquiry, designated the mythical Legend by him before. The legacy problems of the Legend, as we discussed in Chaps. 1, 2 and 3, also carry over to the way science and its elites interact with society. This I will address in the next chapters.

Before doing so, I will briefly discuss Ian Hacking's general thoughts about pragmatism mainly based on his contribution to Misak's *New Pragmatists* (Misak, 2007). Hacking (1936), who I have cited frequently already, has contributed significantly to the philosophy of science. This not from within one particular school of thought, but always taking his own point of view and critically reflecting on the thoughts of others and himself. Hacking is a truly independent thinker who kept his intellectual distance to the logical positivists of the Vienna Circle, to those who critiqued the positivists, to most of the new pragmatists and to pragmatism. He refuses to be labelled. He is however not a nihilist, nor a plain relativist or sceptic and he is not at all in total doubt about science. Like many of the philosophers we mentioned until now, he started in physics and mathematics and then changed to philosophy in England in the 1950s. He thus escaped from the omnipresent formative influence of logical positivism in the USA in those days and confesses that the work of Popper has been his main influence, which lead Hacking to Peirce. With that philosophical background, he came to the USA in 1974. He says he had no idea why so many young American philosophers found Rorty's *Philosophy and the Mirror of Nature* (Rorty, 1979) so exciting' (p35). That book in fact was, Rorty's very progressive rejection of positivism and rediscovery of pragmatism which for Hacking coming from the UK was not a surprise, but in these days for mainstream American philosophy it surely was. Hacking in his typical argumentative style refuses to be regarded a new pragmatist, but on several of the main themes agrees with the pragmatists: the idea that knowledge has no timeless foundation and is fallible, but that *'science has the unusual virtue being intrinsically self-correcting'*. That came to him via Lakatos and Popper but he says: *'it never occurred to me that all knowledge needs foundations, so I did not well, understand what Popper opposed'*. *'Frege had a dream of understanding a pre-given truth that made arithmetic certain, but I never caught the dream'* (p36). *'When I was a student, the search for certainty seemed as dated as Edwardian clothing soon to be favoured by Teddy boys.'* In this context he agrees very much with Dewey *'scathing phrase (of the) 'spectator theory of knowledge*' which occupied the classical analytic philosophers. He agrees with the pragmatist's idea, first proposed by Peirce of the community of inquirers that is instrumental in achieving and testing accepted beliefs. He applauds the pragmatists 'for taking a look' at the practice of inquiry in that sense he says he has always been a pragmatist looking at *'real-life examples and real-life expertise'*. He argues however that this *'is now no more characteristic to pragmatism than it is to any other contemporary style of philosophizing'*. Hacking as quoted in Chap. 2 before said that *'his view (that) realism (of theories and claims) is more a matter of intervention in the world than of representation on words and thought, surely owes much to Dewey'. I recognised that Dewey has there been before me. How did I get there? By talking to my scientific friends'* (p41). In a very interesting paragraph about the problem of the reality of non-observable theoretical entities in physics, he says that for physics it does not make the slightest difference. *'Perhaps it does matter to the funding of physics: It was once alleged that in the journal <u>Nature</u> that the fallibilism of anti-realism of Popper, Kuhn, Lakatos and Feyerabend caused Mrs*

Thatcher to put a spoke in the wheel of British physics.' Actually, Hacking argues she wanted economic returns on investment, *'cash value and saleable results.'* (p42) After much praise for Dewey, James and Peirce and after he connected Nelson Goodman and Erwin Goffman to pragmatism, at the end of the only 14 pages of his contribution, he remembers how Goodman liked his review of Latour and Woolgar, *Laboratory Life* written nine (!) years after its publication (Hacking, 1988).

> Latour and Woolgar and especially Bruno Latour have fundamentally changed the discourse about science by 'taking a very serious look' indeed for two years in the 1970s at a biochemistry lab headed by Roger Guillemin. This happened to be the Salk Institute at La Jolla, where I had just in the spring of 1984 visited a Dutch friend who did his post-doc there with another famous group leader. I discovered *Laboratory Life,* a truly seminal book in the summer of 1984 when I was finishing my PhD on the immunology of human leukaemia's and had already started to work on HIV/AIDS. Kuhn's book was absolutely an eye opener, but this approach of 'taking at good look to science' made a lot of sense to me while culturing cells, doing assays and playing the international games of conferences, publishing and the first experience with grantsmanship. Yet, it was totally different from anything I had read before (Miedema, 2012) (p15).

4.4 Beyond the Legend

In pragmatism and its view of scientific inquiry, the 'external' criteria and values do come in, at the stage of testing of reliability and *robustness* from the perspective of the potential user and stakeholder in society (p194) (Kitcher, 2001). This *invasion* of science by external values and perspectives of societal stakeholders, politics, governments and the diverse publics with their interests and problems is felt by many as a breach of Enlightenment, the 'modern Cartesian' ideas of rationality of science and investigation. Some of the new pragmatists, especially Rorty has gone to the extreme of this post-modernism with the apparent conclusion that it all comes down to *'having a conversation'* (Rorty, 1979). Science in that view had no special claim on knowledge and is just a matter of politics and debates, where power and interest, money, emotions and vulgar democracy are at play. Unfortunately, but understandably this latter ('post-modernist') interpretation has in the 1990s led to a vigorous discussion between those who got carried away by it and defenders of science. The defenders presented thoughtful reactions, as outlined in Chap. 2 and above, regarding the status of the knowledge claims of science without seeking refuge in unrealistic theoretical epistemologies, metaphysics or plain ideology of the Legend (Putnam, 1981, 1995, 2002; Putnam & Conant, 1990, 1994) (Hacking, 1983, 1999) (Longino, 2002) (Bernstein, 1983) (Haack, 2003; Kitcher, 2001, 2012).

Indeed, as we saw, some of the defenders, like Perutz leaned heavily on these empirical positivistic myths of the Legend in their sometimes, resentful writing aimed at 'post-modernist' writers (Perutz, 1995).

As we have seen in Chap. 2 and 4, philosophy, sociology and history of modern science have over the past 40 years converged to a more pragmatic naturalistic view of science. This is based on the study of the <u>practice of science</u> by sociologists and the so-called new pragmatists and scholars that are close to the philosophies of pragmatism but also by more independent scholars. The conclusion is that scientific knowledge is robust and trustworthy, not because science applies a unique method, with formal rules founded on a metaphysical framework, that provides an algorithm for arriving at 'truth'. It holds the view that the way knowledge is produced in science is based on a way of working that is very robust and rests on continuous collective inquiry and intersubjective testing, to decide again and again what are the best insights, theories or beliefs. Testing, validation and failing, involves the testing of knowledge claims in the various theoretical and historical contexts, but including that of the practice of the corresponding problems in the real world. Knowledge has significant value if it proves to be useful, for peers, or if it is more than a consistent theory and whether it also successfully informs our actions in the real world. This in fact relates to science, all sciences and research, as it is how researchers in and outside academia and universities have come to know it by doing it – and not in the least, because it is how it is actually very successfully being done and has successfully been done since centuries.

Having shown the power of pragmatism in explaining how knowledge is produced and how in open two-sided interaction and communication with the relevant publics the impact of inquiry on the real world and our social life could be improved, one wonders why it has not become mainstream philosophy and sociology of science. Why has it gone through a decline after 1945 until at least the late 1960s. I have in Chap. 2 discussed the epistemological side of this coin, in the philosophy and sociology of science the Legend was too strong to be replaced, or better said, to let is self be replaced. From the 1920s on, Dewey's influence and the influence of pragmatism in particular on the educational system, liberalism and political thinking in the USA cannot be underestimated. After 1945, in that domain of society, an additional and possibly even more important debate with political and religious factions in the USA had to be fought by Dewey, and by his followers after his dead in 1952. The successful launch of Sputnik by the Soviets in October 1957 was a shock to the USA with major long lasting effects on its politics and its science (Lepore, 2018). Because of Sputnik, researchers engaged even more in basic natural science who since 1945 were already funded largely and generously by the military, and the NSF was told to deliver to the problems and needs of the military and the space race with 'The Reds'.

After 'Sputnik' and with the Cold War getting more and more impact via the political conservative factions, Dewey's liberal and progressive educational method and his humanistic vision of science and society was openly blamed for the lack of educating competent natural ('hard') scientists who could compete with the Soviets. Dewey was accused for not rejecting Soviet communism, especially Stalinism and of anti-religious sympathies. This was, as we saw in Chap. 2, in support of the popular positivist and the Mertonian vision of science. Importantly, it resonated with the fears of the positivist philosophers of science who had before the war emigrated to the USA from Europe for influence from Marxist and dictatorial politics on science. Although the accusations of Dewey where very poorly supported and mostly derived from poor reading of his work, Dewey and pragmatism despite significant and diverse support from Sidney Hook and the more moderate Reinhold Niebuhr were side-lined through political and public pressure groups and also eventually by President Roosevelt himself. In addition, the subsequent wave in the 1970s when positivism was associated with technocratics and warfare, the fact that the prominent initiators of the Frankfurter Schule, Marcuse and Horkheimer erroneously associated pragmatism with positivism and scientism and repression had its effects. This is important with respect to our understanding of the temporary decline of pragmatism and its impact on science and research in the twentieth century. The social and political analysis of this decline is of great interest, since it is a case in point how external influences of political, cultural and religious values and opinions shape the growth of modern science. It is however outside the scope of this book, and I refer for that to Patrick Diggins' detailed history of *The Promise of Pragmatism* (Diggins, 1994).

References

Barker, G., & Kitcher, P. (2013). *Philosophy of science: A new introduction.* Oxford University Press.

Beck, U., Giddens, A., & Lash, S. (1994). *Reflexive modernization: Politics, tradition and aesthetics in the modern social order.* Polity Press.

Bernstein, R. J. (1983). *Beyond objectivism and relativism: Science, hermeneutics, and praxis.* Basil Blackwell.

Bernstein, R. J. (2010). The pragmatic turn. In *Cambridge.* Polity.

Bourdieu, P. (1975). The specificity of the scientific field and the social conditions of the progress of reason. *Information (International Social Science Council), 14*(6), 19–47. https://doi. org/10.1177/053901847501400602

Dewey, J. (1939). *Logic: The theory of inquiry.* George Allen & Unwin.

Dewey, J., & Rogers, M. L. (2016). *The public and its problems: An essay in political inquiry.* Swallow Press.

Diggins, J. P. (1994). *The promise of pragmatism: Modernism and the crisis of knowledge and authority.* University of Chicago Press.

Gibbons, M., Limoges, C., Nowotny, H., Schwartzman, S., Scott, P., & Trow, M. (1994). *The new production of knowledge: The dynamics of science and research in contemporary societies.* SAGE Publications.

Haack, S. (2003). *Defending science–within reason: Between scientism and cynicism.* Prometheus Books.

Habermas, J. (1970). *Toward a rational society.* Heinemann Educational Books.

Habermas, J. (1971). *Knowledge and human interests.* Beacon Press.

Hacking, I. (1983). *Representing and intervening: Introductory topics in the philosophy of natural science.* Cambridge University Press.

Hacking, I. (1988). The participant irrealist at large in the laboratory. *The British Journal for the Philosophy of Science, 39*(3), 277–294. https://doi.org/10.1093/bjps/39.3.277

Hacking, I. (1999). *The social construction of what?* Harvard University Press.

Kitcher, P. (1993). *The advancement of science: Science without legend, objectivity without illusions.* Oxford University Press.

Kitcher, P. (2001). *Science, truth, and democracy*. Oxford University Press.

Kitcher, P. (2011). *The ethical project* (pp. 1 online resource (ix, 422 pages)). Retrieved from JSTOR. Restricted to UCSD IP addresses https://www.jstor.org/stable/10.2307/j.ctt2jbqjz

Kitcher, P. (2012). *Preludes to pragmatism: Toward a reconstruction of philosophy*. Oxford University Press.

Latour, B. (1993). *We have never been modern*. Harvard University Press.

Lepore, J. (2018). *These truths: A history of the United States* (1st ed.). W.W. Norton & Company.

Longino, H. E. (2002). *The fate of knowledge*. Princeton University Press.

Menand, L. (2001). *The metaphysical club*. Flamingo.

Miedema, F. (2012). *Science 3.0. Real science real knowledge*. Amsterdam University Press.

Misak, C. (2013). *The American pragmatistsThe Oxford history of philosophy* (First edition. ed., pp. 1 online resource (xvi, 286 pages)). Retrieved from ProQuest. Restricted to UCSD IP addresses. Limited to one user at a time. Try again later if refused https://ebookcentral.proquest.com/lib/ucsd/detail.action?docID=1132322

Misak, C. J. (Ed.). (2007). *New pragmatists*. Clarendon Press; Oxford University Press.

Misak, C. (2020). *Frank Ramsey: A sheer excess of power*. Oxford University Press.

Nagel, E. (1940). Charles S. Peirce, Pioneer of modern empiricism. *Philosophy of Science, 7*(1), 69–80. https://doi.org/10.1086/286606

Nowotny, H., Scott, P., & Gibbons, M. (2001). *Re-thinking science: Knowledge and the public in an age of uncertainty*. Cambridge Malden, MA, Polity Press

Perutz, M. (1995). The pioneer defended. In *The New York review of books*.

Popper, K. R. (1981). The rationality of scienctific revolutions. In I. Hacking (Ed.), *Scientific Revolutions*. Oxford University Press.

Putnam, H. (1981). *Reason, truth, and history*. Cambridge University Press.

Putnam, H. (1995). *Pragmatism: an open question*. Blackwell.

Putnam, H. (2002). *The collapse of the fact/value dichotomy and other essays*. Harvard University Press.

Putnam, H., & Conant, J. (1990). *Realism with a human face*. Harvard University Press.

Putnam, H., & Conant, J. (1994). *Words and life*. Harvard University Press.

Ravetz, J. R. (1971). *Scientific knowledge and its social problems*. Clarendon Press.

Ravetz, J. R. (2011). Postnormal science and the maturing of the structural contradictions of modern European science. *Futures, 43*(2), 142–148. https://doi.org/10.1016/j.futures.2010.10.002

Rorty, R. (1979). *Philosophy and the mirror of nature*. Princeton University Press.

Ziman, J. M. (1994). *Prometheus bound: Science in a dynamic steady state*. Cambridge University Press.

Ziman, J. M. (2000). *Real science: What it is, and what it means*. Cambridge University Press.

Chapter 5
Science in Social Contexts

Abstract Gradually since 1990 a growing number of critical analyses from within science have been published of how science was organized as a system and discussing its problems, despite, or paradoxically because the growing size of its endeavour and its growing yearly output. Because of lack of openness with regards to sharing results of research, such as publications and data but in fact of all sorts of other products, science is felt by many to be disappointing with respect to its societal impact, its contribution to the major problems humanity is facing in the current times. With the financial crisis, in analogy, also the crisis of the academic system as described in Chap. 3 was exposed and it seemed that similar systemic neoliberal economic mechanisms operated in these at first sight seemingly different industries. Most of these critiques appeared with increasing frequency since 2014 in formal scientific magazines, social media and with impact reached the leadership of universities, government and funders. This raised awareness and support for the development of new ways of doing science, mostly intuitively and implicitly, but sometimes explicitly motivated by pragmatism aiming for societal progress and contribution to the good life.

To get to this next level we need the critical reflection on the practice of science as done in previous chapters in order to make systemic changes to several critical parts of the knowledge production chain. I will discuss the different analyses of interactions between science and society, in the social and political contexts with publics and politics that show where and how we could improve. The opening up of science and academia in matters of problem choice, data sharing and evaluation of research together with stakeholders from outside academia will help to increase the impact of science on society. It ideally should promote equality, inclusion and diversity of the research agendas. This, I will argue requires an Open Society with Deweyan democracy and safe spaces for deliberations where a diversity of publics and their problems can be heard. In this transition we have to pay close and continuous attention to the many effects of power executed by agents in society and science that we know can distort these 'ideal deliberations' and undermine the ethics of these communications and possibly threaten the autonomy and freedom of research.

© The Author(s) 2022
F. Miedema, *Open Science: the Very Idea*,
https://doi.org/10.1007/978-94-024-2115-6_5

In Chaps. 2 and 3, I discussed the current state of science and the underlying assumptions and images of science. I have shown how this has determined the mainstream 'idea of science and scholarship' and how this has distorted the practice of scientific inquiry and academic culture. It was discussed in Chap. 3 how this still has major impact on science and on the community of scientists. In this Chapter, I will focus on how it disturbed the external relationship between science and society. In the previous chapter I have argued for a more realistic vision of scientific inquiry, beyond positivism and empiricism, as found in pragmatism. Pragmatism, I argue, may help to reshape science and the practice of inquiry to restore the practice of science and importantly its relationship with society and increase in a meaningful way its impact on our social life. The idea of inquiry in pragmatism's theory of inquiry is 'outside in'. Research starts with a problem in social life or something the scientists assume lacks proper understanding and is a cause of uncertainty. It is concluded that the problem relevant for science and or social action based on new knowledge. As a result, knowledge claims are produced that are tested in the contexts where the problem of inquiry surfaced. In this chapter I will from this perspective discuss the current ideas about the relationship between the inside and the outside; science and experts and the relevant publics in societal contexts. I will describe some very recent initiatives aimed at novel, or sometimes rediscovered methods to organize science in academia to improve impact. First let's look at critical thinkers and social experiments in the field of Science and Society that have walked these roads before.

I have, in Chap. 1, discussed the critical reassessment of science mainly by politics in the late 1960s, that one may assume, has resulted in the first serious wave of Science and Society after WW 2, that lasted some twenty years between 1960 and 1980. Inspired by the critical social science theorists of the Frankfurter Schule, Marcuse and Horkheimer, our thinking about the interactions between science and society went through a next phase of 'critical theory' in Europe. Dominant thinkers were Habermas, Foucault and Bourdieu and later Giddens, Beck, Lash, Barnes, Edge. They were highly critical about the role of science in society for different theoretical or socio-political reasons. Some argued against the technocratic dominance with its alienating and distorting social effects (Marcuse, Foucault, Habermas, Toulmin, Illich, Beck, Giddens). Some, from a neo-Marxist but also social-democratic perspective, pointed out that not only government with its military interests, but increasingly multinationals had taken over science and that science should be regained and redirected to be an emancipatory force in society (Marcuse, Habermas, Rose and Rose). This movement of 'humanizing modernity' as Toulmin did describe it in 1990 (Toulmin, 1990), questioned the practice of science, its self-image and with it the ideological dichotomy between the 'hard' rational sciences and the 'soft' social sciences and humanities which also juxtaposed 'timeless, abstract, universal, context free' against 'practical, local, transitory and context bound':

> ...the issues at stake were broached during the 1960s and 70s, in a public debate about the
> aims of higher education and academic research. The debate was dominated by two vogue
> words: on one side "excellence", on the other side, "relevance". The spokesman for

"excellence" saw institutions for higher learning as conserving the traditional wisdom and techniques of our forefathers, while adding to the corpus of knowledge. The focus was on the values of established disciplines....: the subjects should keep their intellectual instruments polished and sharpened....at all cost preserving their existing merits. The spokesman for "relevance" saw matters differently. In their view it was not valuable to keep our knowledge oiled, clean and sharpened, but stored away: it was more important to find ways of putting it to work for human good. From this standpoint, the universities should attack the practical problems of humanity: if established disciplines served as obstacles in this enterprise, new interdisciplinary styles of work were needed... The inherited corpus of knowledge was no doubt excellent in its way, but academics in the 1970s could no longer afford to behave like Mandarins (Toulmin, 1990)(p184,185).

In these days the call for societal relevance of academic research was strong and many university academics were visibly active in public and political debates. This shift was indeed also seen in the research agenda of academia as described very insightful by Toulmin (Toulmin, 1977). He wrote: from *'the focus on disciplinary autonomy and excellence and the pursuit of pure knowledge and technical refinement', From 'Leave us alone to do our own academic thing. Take away your concrete interdisciplinary problems, to knowledge focused on problems and issues that are relevant for human applications'.*

Free Chemistry

After the twenty years of economic growth and prosperity after WW2, in the sixties a widely felt threat and danger of the Cold War, of global nuclear war and the war in Vietnam was felt. The war in Vietnam, which from at least 1967 determined daily prime-time radio and TV evening news in the US and Europe was a dominant divisive political issue, also at the dinner table in my family home. Footage from the battle fields on a daily basis are considered catalysers of inducing a broader disappointment and distrust in the younger generation of the role of science and technology in society. The new generations had not experienced the effects of the war or the poverty of the Great Depression and experienced liberty and freedom to make up their own mind, less dependent on the 'old politics' and sociocultural ideas dominated by religion. It was a mix of worries about pollution and environmental threats expressed in the works of Rachel Carson and The Club of Rome, and in the seventies of recession and gloomy socioeconomics.

For this change of our appreciation of science therefore, historians and sociologists of science point to the cultural and political developments in the 1960s. The historic anti-establishment movements of the summer of 1968, mainly from students in the US, France, Germany and also in some other countries in Europe were however short-lived (Miller, 1994). Still, for twenty years they have had a significant effect on science and its relationship with society. Science was seen as the main power in society that could do harm but when tuned to the needs of society could do a lot of good. Famous initiatives

(continued)

were the 'science shops' in The Netherlands and a unique nationwide public debate in our country in the late seventies on nuclear energy in which many of my friends actively participated. When I entered university in 1971, vigorous debates went on about the role of science in society, social responsibility of scientists and who controlled the curriculum. Friends of mine after their B.Sc. in chemistry went on doing a M.Sc in Free Chemistry in Groningen, a mix of chemistry, science studies, social theory and sociology and easily found interesting jobs in these fields after graduation.

Several mostly local and national movements in the 1960s and 70s, both in academia as in the political domain, have responded to the disconnect between science and society. Science and Society and later Science and Technology Studies (STS) became an academic trans-discipline in the late 1970s with its critical stance and appeal for responsible science and societal relevance. The movement inspired the idea of legendary Science Shops, and many other quite different forms of public participation, public hearings, problem-driven bottom-up movements where citizens and lay publics could meet with academic experts for help, advice but also influencing and building joint research agendas. In many countries, academics became organized to become politically and socially active. Conceptually, this in some respect developed in parallel with the development and critique of the popular image of science described in Chap. 2. Studies from sociologists, political theorists, and from the then newly established field of STS about positive and negative interactions between science and society provided insights for these actions (Ravetz, Blume, Rose and Rose, Habermas, Sarewitz, Guston, Bijker, Rip, Meulen). These analyses, and actions, have led to many small-scale local actions and interventions to engage and increase societal relevance in the practice of research. Despite all this, these movements from outside and inside universities have not changed the practice of mainstream academic science in the longer run.

COVID-19: the public looks on and talks back.

As I am writing this, March 30, 2020, we are in the first surge of the Corona Crisis, the COVID-19 pandemic. In times of war and crises like the corona crisis the dangers and pressures are such that the response from government goes beyond partisan lines, one would think. Not always. Anthony Fauci and Deborah Birx have just yesterday convinced Donald Trump that the virus is not a hoax of the Democrats. That it really is a very serious health problem, with a high death toll to be expected even when the US government in close collaboration with experts in the public health system responds adequately. Experts these days are talking with the responsible politicians, are on news

(continued)

and talks shows everywhere you look. The people after some time in large majority accept their advice, no matter how disruptive to social life and economy. When this crisis develops, it is being asked why 'we' have not invested more in health care, public health intelligence and research and why there are no facilities and institutions who can ramp up to meet the scale of this pandemic. Indeed, we stare at the screens and start to reflect on who and on which grounds we are making the choices for all kind of things of science and technology and how that shapes our social life. This COVID-19 problem is a threat so immense and as its effects are highly visible on the evening news, that there is unanimity regarding expert opinion. In response politics and publics ask from science: 'Screening, testing, treatment, therapy and a vaccine, now'. Dealing with uncertainty and its resulting insecurity about the course of the pandemic is unbearable. It happens that in some daily discussions in the media non-experts denounce the experts for lack of certainty and adjustments they make in their science analysis and advice. Risk of disease and death, in our times are unacceptable. Experts in the field of infectious diseases, however, know uncertainty from experience and from the recent history of pandemics, despite their excellent modelling based on high quality mathematics and sophisticated biology. They are openly and honestly declaring that there are many critical unknowns and new data keep coming also to them, The public, in parallel, via the media see on a daily basis new data coming that is immediately before their eyes and used by scientists to update the models which changes the predictions and informs policy. This is hypermodernity. The scientists study the virology and public health but also social and economic disruption and are weighing the evidence. The publics in the meantime, with a feeling to be subjects of the study, are aware and asked to adapt their behaviour to influence reality. Researchers, in fact are mediators between science and politics, like Fauci, and politicians are deliberating every day to come to the best policies to deal with the pandemic. Weighing health risks, economical risks and social disruption, the politicians in the end have to decide. This functions best in democracies when free flow of information, communication and undistorted discourse is possible, which even in modern democracies is not obvious, as we have seen not only in the White House Corona briefings over the past months.

5.1   The 'Pragmatic Model' in Frankfurt

Most scholars in the philosophy of science and political theory recognize Habermas to be the most important link between American Pragmatism and European Continental philosophy. As discussed in Chap. 3, Habermas in his *Knowledge and Human Interests* discussed at length the work of Peirce on the logic of inquiry (Habermas, 1971). This is emphasized by Habermas in the Appendix which is his inaugural lecture of June 1965. He explicitly endorsed the essence of pragmatism in the relation with real world problems and the values that in inquiry do come with them and the role of the community of inquirers in the process of defining acceptable knowledge claims. In *Technik und Wissenshaft als Ideology* (Habermas, 1968, 1970) that I bought in June 1976, he describes the penetration or in his words, the rationalization of the social sphere by science and technology in our modern late capitalist Western societies. Science refers in this context to the natural sciences with their positivistic philosophy. The classical separation, he argues, between science and its knowledge and the social life in society does not exist anymore and this has two results. The sciences are coupled to and are drivers of economic and technologic innovation shaping and dominating our social life. At the same time, they became uncoupled from the humanities, 'from the humanistic culture', with which reflection on its practice is lost (p55) (Habermas, 1970). Habermas argues that the capacity to control nature and social life are assets of science, which has allowed for at least the most of us, to live a better life and in comfort, but that capacity has become the problem. The institutionalized rationalization that comes with science and technology, is largely uncoupled from the needs and problems of the diverse publics and has become dominant and repressive. The logic of science and technology as a power, he argues, penetrates society and politics, has its own intrinsic dynamics, which brings the problems in social life as we see them unfold. Habermas proposes that science and the publics should work on their 'self-understanding' in order, citing Dewey, to be able to come to a *'pragmatistic model'* that is associated with democracy in which *'the strict separation between the function of the expert and the politician is replaced by a critical interaction. This interaction not only strips the ideological supported exercise of power of an unreliable bias of legitimation but makes it accessible as a whole to scientifically informed discussion, thereby usually changing it'* (p66, 67) (Habermas, 1970). This model allows for social interests and their value systems to be played out in the deliberations and allows for true legitimation of policies before the public. He continues discussing Dewey:

> For Dewey it seemed self-evident that the relation of reciprocal guidance and enlightenment between the production of techniques and strategies on the one hand and the value-orientations of interested groups on the other could be realized within the unquestionable horizon of common sense and an uncomplicated public realm. But the structural change in the bourgeois public realm would have demonstrated the naïveté of this view even if it were not already invalidated by the internal developments of the sciences p69.
>
> He refers specifically to the confusion of *'the actual difficulty of effecting permanent communication between science and public opinion with the violation of logical and methodological rules. True, as it stands the pragmatic model cannot be applied to political decision making in modern mass democracy. The reason is….the model neglects the specific logical characteristics and the social preconditions for reliable translation of scientific information into the ordinary language of practice and inversely for translation from context of practical questions back into specialized language of technical and strategic recommendations.'* (p70).

He argues that in USA politics since the war this is being practiced and describes the necessary sequences of actions in such a case. He argues for a long-term science policy which *'attempts to bring under control the traditional, fortuitous unplanned relations between technical progress and the social life-word.'* (p72). He, with Dewey, is thus well aware that the ideal conditions for this pragmatic model are generally not present. Habermas, as Dewey and the pragmatists, is neither a nihilist nor hard-boiled sceptic paralysed by the idea that everything is determined and defined by power games and by unchangeable practices of repression and domination. Importantly, Habermas, despite coming from a Marxist tradition of political theory, and being the successor of Marcuse and Horkheimer at the Frankfurter Schule, does not see human values and interests per definition as distortive forces in the interaction between science and politics. It is the belief in human agency and the trust that communication and ethical discourse is possible and wanted by most, but they have to be consciously, monitored, regulated and managed in well-designed and carefully executed open democratic processes. As in most of his later works of political theory, communication, mediation and discourse ethics, freedom from domination and repression, is essential in all phases of societal development and social action where these deliberations need to take place. In his emphasis on communication Habermas builds heavily on the work of the sociologist George Herbert Mead a prominent pragmatist in the early decade of the previous century.

Finally, citing D.J. de Solla Price's now famous studies published a few years before, (Price, 1963) he mentions the problems of specialization and barriers between disciplines and scientific communication with the overwhelming numbers of articles and journals and issues of military research and secrecy. He describes the requirements of political and institutional advisory bodies, societal organization and the organization of the research process, that will facilitate the model. Remember, these were the days when environmental issues, nuclear energy, radioactive waste and the nuclear arms race, the first signs of the energy crisis and a war in Vietnam for which the motives and logic had long evaporated, were the topics of major public concern, debate and protests. In this technocracy the publics felt alienated in all kind of respects as they felt that their issues and concerns were not being dealt with. These were seen as the consequences of blind belief in and application of the natural sciences not only to the war in Vietnam which aroused massive political movements and the student protests of 1968. In this stage of capitalist society, the old Marxist materialistic dialectics of 'capital and proletariat' had lost bite. Because of the atrocities of the Stalin regime that were generally acknowledged and condemned and because synergy of science and capitalism has brought enormous economic welfare, at least in the West. The discussion was whether science and technology are in the true humanistic meaning of the word being used to relieve hardship and inequality and promote 'the good life'. It seemed that science and technology that claimed to be neutral, was exploited by commercial and military interests that were not effectively being controlled by public deliberation in our democracies. It has been argued that al lot of research can be initiated without the involvement and mediation of governmental agencies when a public focussed around a well-articulated problem talks with researchers. This can be at the national level but is increasingly happening at regional levels when networks have been established of citizens and representations of science and research around major themes like public health, welfare or environmental policy. This problem of the relation between experts, the public and politics thus concerns the democratic way to set the research agenda and how expert scientific knowledge is taking into account in the formation of national and local governmental policies. Leaving out the (neo-Marxist) jargon and undertones, this piece could after more than 50 years easily be mistaken to be part of a paper on Open Science in our time and age pointing to the Sustainable Development Goals of the United Nations, a list of societal issues that is in number and calamitous impacts not less compared to the list of 1968.

A Fellow Traveller in Science in Transition
The initiators of Science in Transition, as described in Chap. 3, had very different experiences and diverse perspectives on science and academia. Introducing one of them, Huub Dijstelbloem, in this particular chapter is appropriate and of relevance. Dijstelbloem, is since 2009 a Senior Researcher at the Dutch Scientific Council for Government Policy (WRR) and since 2015 also appointed professor of philosophy at University of Amsterdam. Before that he was a Program Coordinator at the Rathenau Institute, the institute of the Royal Academy in The Hague which advises government policy on science, technology and innovation. I didn't know Huub, but his straightforward contribution to the Spui25 debate of October 2012 was in agreement with my own views about science. We briefly met that evening after the debate and decided that we should keep in touch in order to jointly prepare for a public action. The end of November 2012, over a simple dinner in 'The Ysbreeker' overlooking the Amstel, we exchanged our views and ideas and started to talk about a plan. Our professional backgrounds were quite different, but we had a very important common interest, namely the problematic interaction between science and society. We had both come to the conclusion that there were two problems there: how science was organized and how its communication with politics and publics was thought of and organized. Just the year before he had edited (Dijstelbloem, 2011) with Rob Hagendijk a very nice book about trust in science with modern philosophical and sociological perspectives. When we started, I discovered his work on research evaluation done with Jack Spaapen (KNAW) which obviously was most relevant to our later discussions on Incentives and Rewards (Spaapen et al., 2007). The current chapter deals with one of the major topics that Science in Transition believed should be addressed to improve the practice and impact of science: science in the societal context. This happens to be the topic of Huub Dijstelbloem's appointment with the Dutch Scientific Council for Government Policy (WRR) where he contributed his views to many advices and reports. His PhD thesis published in 2007 and much of his work since 2007 is about the science and society interface, mainly about policy advice. In his thesis he discusses the problems of the interaction and deliberation between scientific experts, representatives from the public and politics. He takes pragmatism as his main conceptual perspective, starting from Dewey's *The Public and its Problems*, (Dewey & Rogers, 2012),

(continued)

Latour's work on Pasteur (Latour, 1988) and the work of Habermas discussed in this chapter. In his thesis he analysed in great detail the fascinating case of the response of the government and public institutes to the HIV epidemic in 1982 in the Netherlands (Dijstelbloem, 2014a, b). He described the case in which the blood supply foundation asked the gay community to voluntarily refrain from donating blood, with major roles for Vincent Eijsvoogel and Pim van Aken, directors of CLB Blood Transfusion Service at Amsterdam (now Sanquin), Roel Coutinho the director of the Municipal Health Service of Amsterdam and representatives of the community of gay men. This for me is special, as I was in these days doing research at CLB and knew the issue and nearly all the actors in this significant example of boundary work between science and the public.

5.2 The Problem of Power

The ideas of Dewey and also Habermas about deliberations between experts and representatives from the public and politics have by many scholars been seriously criticized for being naïve with respect to the distorting effects of all kinds of power and on too much reliance on the proper functioning of formal institutions. Although both men were aware of these distorting forces, they believed that in principle people are to be considered moral beings that want to achieve the conditions that allow for the betterment of social life, 'the good life'. Habermas has his whole life, literally until this day, worked to develop this concept of 'discourse ethics' and 'communicative reason' which provides the intersubjective foundation for actions of people and institutions. Dewey emphasized social and personal education to emancipate the public and to provide them with the means to engage in civil society. This was not as much a believe, but for Habermas a moral principle on which most people build, when they engage in social and political life. This communication, in contrast to dialectics and conflict was also the main theme in the work of George Herbert Mead (1863–1931) another major American pragmatist who was an inspiration to Habermas. In this perspective communication and language are powerful instruments to deal with subjectivity, to achieve intersubjectivity as a form of objectivity and importantly to expose the misuse of power in pursuance of particular interests. Here we see the concept of 'objectivity' making the same turn in social theory that in pragmatist epistemology turned from individual to 'intersubjective'. Habermas disagreed with the philosophy and writings of Foucault, Nietzsche and others, in which the corruption and domination by language and communication is central in how power is executed and totally penetrates social life. Foucault and others analysed and exposed the distortions of inequalities of power and the adverse and perverse effects of misuse of power in many major sectors and institutions of society like medicine, sexuality, education, law in which discipline and punishment are practiced from without positions of power using language and communication to achieve to discipline. With Nietzsche in the background, these analyses imply that it is hard to image how to curb these perverse effects of power and thus hard to avoid scepticism and nihilism.

Bernt Flyvbjerg in *Making Social Science Matter* (Flyvbjerg, 2001) criticizes the social sciences from within and starts from the perspective that power is pervasive in social life and politics and has to be dealt with when we are thinking about communication as driver of social action in the public sphere. Flyvbjerg, inspired by Bourdieu and Latour, writes about social theory and he also empirically studied social actions at local levels of citizenship and governments. In these contexts, he sees values and power as prominent factors in debates. He concludes, correctly, that Foucault and Habermas are both very aware of the problem of power but approach them differently and complementary. Habermas indeed argues for engaging with publics focused around problems but believes this must be taken up by the institutions and change the relevant institutions and agencies. Foucault sees institutions often as part of the problem, because they inevitably will define their own goals and agenda. He believes that issues of power will have to analysed, understood and dealt with in the specific contexts where they occur. Habermas, with Dewey, clearly tries to avoid and anticipate situations of conflict through proper communication and understanding of all sides, but Foucault of course interprets the evasion of conflict as suppression and restriction of freedom. In agreement with other scholars, Flyvbjerg sees conflict not as a danger per se, but believes that it can result in new opportunities and change. Hence if one engages about controversial issues of social action with parties with different interests, one has to choose, dependent on the context of the issues, which of these approaches to employ. Obviously, this also depends on the level of democracy of governance at the regional or national level where you are. It is of interest to read in exactly this context Diggins account how the American framers of the Constitution were not naïve romantics either, regarding the problem of power, private interests and the abuse of language. They 'followed Locke and Hume but not Descartes and Kant' and anticipated conflicts and saw to it that in the words of their criticizers *'they* (had been) *burdening the young Republic with excessive reliance on controlling mechanisms, such as the separation of power, instead of centralizing all authority in a single national assembly that would represent a virtuous citizenry'* (p428–434) (Diggins, 1994).

5.3   Well-Ordered Science

The problem of power and distortion in philosophy and theory about 'ideal deliberations' in a 'well-ordered society' are well known (Rawls, 1999). These theories about justice in society are to be read and used as aspiration and guidance for our thinking about how to act in social life. Philip Kitcher (Kitcher, 2001, 2011) has in that vein, proposed a theory for 'well-ordered science' where in a democratic fashion the agenda for scientific research is being set in order that society will have optimal benefit of the research. This idea has been developed and discussed in detail in two books published ten years apart (Kitcher, 2001, 2011). He discussed all the issues that are related to allowing external voices and opinions in the deliberations about science and the science agenda. Although Kitcher expresses his doubt and anticipates the critical opinions of the majority of scientists, he came to the conclusion that we must somehow engage in this. He concluded that science based on the Legend was misguided and that we should aim for 'significant knowledge'. Significant knowledge for Kitcher being knowledge that has impact directly or indirectly on real world problems. He is clear about the purpose of science: *'Even the slightest sympathy with pragmatism (in either the philosophical or the everyday sense) will recognize circumstances in which esoteric interests of scientific specialists ought to give way to urgent needs of people who live in poverty and squalor'*(p110).*(Kitcher*, 2011).

Kitcher uses the term 'significant knowledge' for the type of results of inquiry which contrast with the 'esoteric' form of knowledge production. It is obvious that here problem choice in inquiry and values next to cognitive criteria to steer that process are central. This determines the quality of inquiry in terms of its potential contribution to the body of knowledge and to decide and structure social or political action in the context of problems and needs. Kitcher does analyse the situation of the research agenda in the institutions. He concludes that in the past one hundred years, research has shifted from private to public but there is no oversight of the research agenda, either at the institutional nor at the national level. It simply is a list of a series of actions 'any institution of public knowledge' has to do. This poses the question which problems to study based on their estimated significance (p101) (Kitcher, 2011) It is unclear at exactly what level the institution here referred to, should act. Kitcher as many others seems to consider this as a 'black box' or a product of the legendary 'invisible hand' and states *'Science has evolved by happenstance'*. He does however mention with admiration the intervention by Vannevar Bush' who with very visible hand *'brilliantly developed a utilitarian case for public support of science' 'but preserves the idea of scientific autonomy; the public is to provide, but the community of scientists is to decide…'*. *'Optimistic visions like those…contrast with others that view any system of public knowledge as potentially oppressive'* (p101). Here he refers to Foucault who he says, despite his rhetoric, produced real insight in this problem, not citing many other influential scholars of Critical Theory as for instance Habermas. Fortunately, Kitcher explicitly discusses issues of power and interests and believes that *'they should not be used to scoff at philosophical ideals on the grounds that they require a lot of changes.'* He mentions major aspects of the current practice of science that are obstructive or even do run counter to the ideal of well-ordered science which require several changes including: competition in academia; flaws of vulgar democracy infested in public engagement; privatization of university research; neglect of many publics and their problems in the less affluent parts of the world; myopia on the part of academics in problem choice options.

The problem of majority vote on issues where expert opinion is of great importance and the majority may not be well informed or able to come to a justified opinion has been bothering Kitcher. He coined the very ominous terms *'vulgar democracy'* and *'tyranny of the ignorant'* describing all the fears and nightmares of not only elitist scientists when they have to consider the idea that even well-informed and educated citizens get involved in decision making about the different agendas for inquiry or science driven policy making. Kitcher also has these fears and proposed a form of *'enlightened democracy'* in his earlier works to mitigate these threats (Kitcher, 1993). He considered the problem of elitism, but came to the conclusion that experts, who are to be trusted to be able to properly understand and make judgement, should inform groups of citizens and the experts then should decide based on the various perspectives (p133–135).(Kitcher, 2011).

Kitcher does not discuss, or only very indirectly discussed, one very obvious 'visible hand' that already for ages is steering the national research agenda and the agenda of institutions, namely how economy shapes science (Stephan, 2012), funding policies and money from whatever source available to institutions, individuals or groups of researchers determine to a major degree the agenda. We have seen in Chap. 3 how the incentive and reward system, has developed into a distorted system and how it determines our problem choice and the research agenda and the more strategic choices made on daily basis by committees of researchers all around the globe. The research agenda of funders has in most cases until recently been determined by internal scientific arguments based on quality measures of science, as its scientific committees are preferably populated by elite scientists.

In the final pages on Well-Ordered Science (p131–137) (Kitcher, 2011) he discusses moral issues, rephrased here by me: Are scientists obliged to work on the research that will yield the most significant knowledge? Should this be organized by procedures, so they are made to do the research that is ethically required of them? This is, he says, of course not what we should want. When scientists with their goals and preferences take part in ideal deliberations, they express their motivations and will be heard. There are, however, situations of emergency, Kitcher correctly points out, when there will be overriding public demands and researchers must drop their work and join to work on major problems. Most famous, is the Los Alamos Project and other major research projects during WW2, but think of pandemics of Flu, HIV, microbial warfare, the financial crisis of 2008. At the time I am literally writing these lines, the COVID-19 pandemic is such a global emergency and we see that scientist, in international networks from many disciplines have started to work together sharing data, materials and concepts in order to limit the damage to individual and public health and subsequently to try to avoid as much as possible the ensuing economic depression and its dramatic social effects.

Public representation is a problem for most writers on deliberation and the interaction between experts and the public it involves. This is most prominent in issues of political choice when they have complex technical or scientific components and the choice between policy options involves scientific advice and expertise next to social-economic and political arguments and values. Kitcher is very honest about the limitations of his philosophical position: *'My original thoughts about well-ordered science and the potential of groups of citizens to participate in deliberations that are simultaneously broadly representative and well-informed were advanced in ignorance of the actual experiments that have been carried out.'* (p223) *(Kitcher,* 2011*)* Based on the two cases he discussed, he concludes rather gloomy that one of the greatest stumbling blocks is loss of authority and trust of experts. He writes: 'the *situation of our democratic discussion is currently so dire that no redress along the lines proposed is possible: there will always be loud voices decrying any efforts to rebuild trust in expertise'*. He concludes with the observation that only if the majority becomes aware and we start to address this problem, deliberative democracy will have a chance.(p226) (Kitcher, 2011).

5.4 The Legend Meets Reality and Pragmatism

Kitcher, influenced by Dewey and the new pragmatists, has converted since the late 1990s in his thinking about science and inquiry from analytical philosophy to pragmatism. This may explain why he writes about well-ordered science and the interaction between science and society as he does in these two important books cited above. These books are important in my mind exactly because of his philosophical history. He was well aware of the intellectual and emotional struggle of what it was that he had to leave behind and to face the accusations of engaging in relativism, post-modernism and being anti-science. Because of his background it seems he does yet not fully engage with pragmatism on two accounts: the ideas of Dewey, who he otherwise cites as an important inspiration, about the essential engagement with the public and their problems in inquiry and the more theoretical work of Peirce on that same issue. The pragmatic idea that the results of inquiry are really tested for value and acceptability by the community when translated into action, within or outside science depending on the problem they started with, is of great relevance to his idea of significant knowledge and well-ordered science. Mark Brown and Huub Dijstelbloem, and several others have in their essayistic reviews commented on these issues (Brown, 2004, 2013; Dijstelbloem, 2014a, b). A major criticism was that Kitcher did not mention the wealth of studies that were published between 1990 and 2010 on the many cases of public and citizen engagement and about prominent public debates with problematic interactions of scientist/experts and the public. To understand the major aspects of this interaction, Mark Brown in his own work goes from philosophy, sociology, STS to political and social theory (Brown, 2009). He explicitly observes science and society from an integral pragmatist perspective. Dijstelbloem, like Brown, suggest that the idea of 'the public', and more general groups of selected citizens, does not hold, as in many issues of policy making and expert advice, the debates and interactions are between designated groups of citizens who are concerned and directly affected by the respective political, public or private actions in their community. This indeed agrees with Dewey's idea of publics that are focussed and get organized in time and place around well-defined problems. Compared to the established governmental institutions, these publics are dynamic and fluid like their problems are. Dijstelbloem, as an example of such a public, performed a detailed analysis of the history of the initial institutional response in 1982 to the HIV epidemic in The Netherlands (Dijstelbloem, 2014a, b). The most interesting thing was that around that time in the AIDS epidemic, HIV was to yet be discovered. Yet, several affected and concerned parties from society convened through an initiative of CLB, a national not-for profit foundation doing blood research, diagnostic services and a major national producer of blood products, clotting factors and other blood plasma derived medicines. The board of CLB was worried about the unknown pathogen that they anticipated caused AIDS and apparently was transmitted by blood products produced from blood donations from infected donors. To protect patients receiving blood transfusions and other blood products and specifically haemophiliacs who regularly need blood products produced from blood plasma batches involving pooled donations of thousands of donors, the idea was to ask gay men to refrain from donating blood. This directive was discussed before action was taken, with respect for feelings of discrimination on part of the gay community. At the table were the representatives and experts of CLB, representatives from physicians who treated haemophiliacs, representatives from the gay community (Men who have Sex with Men, MSM). The chairman was Roel Coutinho, director of the Municipal Health Service of Amsterdam and involved in a Hepatitis B vaccine study in gay men. The gay men objected as expected, casted doubt on the relevance for the local situation of the scientific arguments mainly based on data from the USA. It took four months to agree on a directive in which not homosexuality, not promiscuity, but having 'multiple sexual relations', which was quantitated to more than five in the previous six months, was the consideration on which one was asked to refrain from donating blood. Interestingly, no representatives of haemophilia patients, government, governmental agencies nor intravenous-drug users had been involved. Apparently, not all relevant concerned publics were aware and organised yet. The process showed that formal institutions need not to be at the centre. The latter contradicts the believe that Dewey's social approach via publics, in contrast to more formal institution approaches, would not have the power required to change and impact policies. This successful activist type of actions by the gay community in the history of AIDS and HIV have been quite common as has been shown by Steven Epstein's *Impure Science* (Epstein, 1996) and are badly needed in many fields of biomedicine, as he showed in his *Inclusion, The Politics of Difference in Biomedical research(Epstein, 2007).*

Dijstelbloem's study and its theoretical interpretation is a nice example of Science and Technology Studies (STS) done since the 1980s. Major researchers in early STS are amongst others Brian Wynne, Wiebe Bijker and Trevor Pinch (see Oudshoorn and Pinch eds, *How Users Matter; The co-construction of users* (Oudshoorn & Pinch, 2003). Sheila Jasanoff has since the 1980s produced an impressive body of in-depth scholarly work on the interaction and relationship between science, science advice and policy making which has guided many researchers since 1990. She is most interested in the dynamics of the policy making process, and I refer to (Jasanoff, 2012), a collection of her papers and especially Chap. 6. In that paper published in 1987 she very concisely discusses how philosophy and sociology of science have revealed the real practice of science and the flaws of the classical image and limited self-understanding of science, which has a major effect on science advice at the boundary when scientists are meeting with representatives and opinions of publics and politics (p103).

Irwin and Wynne's *Misunderstanding Science? The public reconstruction of science and technology* is an excellent series of papers on case studies in which various fields of expertise and its experts are involved (Irwin & Wynne, 1996). Among them is Wynne's famous study on the Sellafield sheep farmers and nuclear fallout in which scientist discovered that the sheep farmers had expertise and ideas of their own which were highly relevant to the problem. The authors thoroughly analysed the cases to understand the problems in the interaction of experts and the lay public. I will not go in detail, but many common issues concerning science and experts become evident when their scientific claims have to face up to public scrutiny, which Irwin and Wynne in quite clear language summarized in their Conclusions (p213–221). They start with their own definition of what Kitcher later termed 'significant knowledge'. They more prosaically call it 'useful knowledge', meaning 'valid and socially legitimate as well as being of immediate practical relevance and use'. Social groups often ignore expert (scientific) knowledge because *'it is not tailored to the needs, constraints and opportunity structures of the social situation into which it has been interjected as authoritative knowledge'*. Experts must be sensitive to *'local contexts and need to listen to and to try to understand user situations and knowledges.'* For social legitimation of expertise it is required to *'reopen ..expert knowledge and its validation all over again - but in more complex, less reductionist circumstances. Often,...the prior context of scientific validation has been shaped by social assumptions and these have been 'black boxed..'.* They latter have not been made explicit and in addition the classical idea is *that validation of expert knowledge is completed before (and insulated from) its social deployment or use'.(p214)* This is as discussed in Chap. 2 in agreement with the scholars who have concluded that in the practice of science 'universal validity' was limited by strict experimental conditions which are hardly ever met in the world outside the laboratory or ideal research setting, which is the reason why despite positive clinical trials in a less ideally selected patient group many medicines fail (Cartwright, 1999).

This is, Irwin and Wynne say: *'the public understanding of science problematique'*, the projection on to *'the public* of *the internal problems and insecurities about legitimation, public identification, and negotiation of science's own identity.* They conclude that this is the heart of the problem: *'all of the troubling experiences of apathy, resistance, plain distortion, and exaggeration which disfigure the public life of science in modern scientific democracy have led to little or no consideration of whether they imply anything might be wrong with the organisation, control, and conduct of 'science' (in addition to its communication).'(p214).* They conclude that the expert's idea of the public is wrong seeing them as a socially amorphous aggregate of individuals with erroneous unchecked assumptions about what the public wants and needs. This is a major point, as argued above and follows Dewey's proposition engaging the relevant citizens, although his work is not cited. Policy issues get a very different, more practical context if representatives of the public that is concerned and affected are involved (Marres, 2007). Several issues relate to the lack of understanding of the specifics of local contexts and their publics. Irwin and

Wynne anticipate that these will be *'uncomfortable conclusions for the scientific community since it suggests a pressing need for debate over the limitations of science as well as its putative benefits. However, in a situation where public groups more often see science as obstacle to development rather than a facilitator, there is little choice.'* (p219). To come to 'more progressive relationships between knowledge and citizenship' they propose 'new institutionalized forms which attempt to deal with these issues' and are sympathetic to small scale experiments and projects from which a lot is to be learned. They realized however that *'such localized and specific initiatives struggle to gain credibility within scientific institutions...*not being seen as belonging *to the preferred and more cloistered world of science..'* They emphasize that it all comes down to the institutions and the organization of science to achieve the required change in the attitude and practice in order to have more impact and with it more social legitimation. We have seen other scholars who analyse the problems of science and come to the same conclusion, although such a strong call for organizational change for instance in the incentive and reward system are rare. In terms of the analysis in Chap. 2 and 4, this sounds like The Legend meets Reality and Pragmatism. Flyvbjerg, (Flyvbjerg, 2001) with his advocacy for *phronesis*, the method of understanding and interpreting in inquiry, accompanies his plea to leave the idea of the Legend behind, as its method that may work for some of the natural sciences, is inadequate for other natural and biomedical sciences and the social sciences. We have seen in the previous chapters how difficult it is to achieve this systemic organizational change, as also in science and academia knowledge, power and interests are entangled (Bourdieu, 1988, 2004; Rouse, 1987, 1996).

ACT UP at the National Institutes of Health, 1990

My blog on interactions with the participants from the Amsterdam Cohort Studies: To confront 21st century challenges, science must rethink its reward system: One of Science in Transition's founders describes how his experience as a young HIV/AIDS researcher convinced him that science needs to change. The Guardian, 12 May 2016. https://www.theguardian.com/science/political-science/2016/may/12/to-confront-21st-century-challenges-science-needs-to-rethink-its-reward-system

'HIV/AIDS research in the early 1980s was a new and exciting field of science. I had started working as a biomedical researcher in Amsterdam, a city with a large and visible gay community. The new disease was a threat to public health and was highly contagious. It was transmitted by sexual contact and in the developed world affected young healthy gay men and recipients of blood and blood products. It took some time to realise that a truly immense and devastating epidemic was going on in sub-Saharan Africa affecting men, women and children. This disease attracted bright scientific minds all over the world, working feverishly to understand the origin and biology of the virus. We wanted to know how the virus moved through the population, entered and killed immune cells and how to counteract it. AIDS patients were dying in the hospitals and we were working as fast as we could towards better therapies for HIV-positive patients. Or were we?

I was very proud when results of experiments from my laboratory were published in prestigious academic journals like *Nature, Science* and *The Lancet*. I felt I had made a significant contribution to understanding and battling HIV. As well as at scientific conferences, we presented our results to participants of the Amsterdam Cohort Studies which started in the late 1980s. These were mainly gay men who helped our research by donating blood samples and filling in lifestyle questionnaires. One evening I presented with my usual enthusiasm new results on how HIV destroyed white blood cells of the immune system. Then a man came to the microphone. "Doctor Miedema, thank you for your interesting talk, but to be honest, it was a bit over my head, with apoptosis, virus particles and what have you. However, what I would like to know from you is whether we should practice safe sex even when my partner and I both are already HIV-positive." I was of course flabbergasted. Here I was with my clever immunological experiments and detailed molecular understanding of the virus, but I couldn't answer this real-world question. And the question made sense. Rephrased in viro-immunological research questions: is it possible and, if so, is it bad to become co-infected with a different virus strain? Can mosaic viruses with increased pathogenicity emerge? We, the smart boys and girls in the lab, hadn't thought of that question. Why not? Because it hadn't come up in the lab. We had informed the patients but forgotten to talk to them, the people we were supposedly working so hard for'. In my 'academic reflex' I translated his question into a research question

(continued)

which could give my team a nice publishable result and a paper. Indeed, not only didn't we talk to the patients to hear their needs, but it was felt that paying too much attention to those needs might be bad for our academic career, unless it would yield top publications. At that moment I realized how detrimental for societal impact the reward system and the corresponding research strategy is when the journal paper is the goal instead of a means to and end of having true societal or clinical impact.'

Public participation is a complex two-way process in which the scientist and experts must reflect on their practice and on needs and motivations of the lay participants. Of course, we must not be naïve and overly optimistic. Complexity reaches the next level as soon as policy making is discussed and the debate is organized tripartite with politician, local or national. There is a host of critical research on how the dynamics of these processes can be gamed by politicians who may well have their own motives and plans which do not a line with that of the public it concerns. It has been shown that in these cases participation is sham democracy and not real deliberation but simple a means to win the public over. (Felt & Fochler, 2010; Wilsdon et al., 2005; Wilsdon & Willis, 2004)

5.5   Rethinking Science

Most authors writing about science, for obvious reasons have evaded the issue, but it is clear we need to re-think the system of inquiry and academia, no matter how hard that may seem to be. A collective of authors lead by Helga Nowotny and Michael Gibbons, however, have just done that and published two remarkable books in 1994 and 2001 (Gibbons et al., 1994; Nowotny et al., 2001). Both books communicate a strong, nearly tangible sense of urgency for change. In 'Re-Thinking Science', they present a thorough comprehensive and dazzling analysis of ongoing parallel developments in society and science driven by socio-economic, scientific and technical, digital innovations that disrupt the way we live with due references to the seminal work of Giddens, Lash and Beck. These developments are the cause of persistent but rapid change that comes with disruption and fundamental uncertainty in society. It affects and changes our basic ideas and concepts about the good life: human interactions, community, communication, identity and belonging, ethics and responsibility, commitment and freedom in the personal, the national and the global public sphere. They point to the blurring of boundaries between science and society, they refer to the 'agora', the many physical but increasingly also virtual regional and national marketplaces where science and society meet and become entangled. Science is invading society, society is 'talking back to science', with at the same time science and research closing the gaps between investigation, action and application.

Giddens, Beck and Lash have argued that it is a logical consequence of post- or hyper-modernity for science to have to be reflexive regarding its own functioning, concepts, results and instruments. Doing sociological research on problems and issues in society, subjects and publics of these studies will immediately be able to know the results and want to become or are engaged. This will affect the social practices and behaviours of the subjects who have an interest in that research. Researchers must deal with this public reflexivity by reflexivity on their part in the practice of their research. This is one of the consequences of modernity, where initially clear boundaries between science and society (church, state, politics) were needed to provide freedom for investigation, in our time of hypermodernity science and social practice are developing and organized in parallel and in continuous interaction. This happens in a common public sphere where their relationship is based on communication, ideally the ethical discourse that Habermas believes is required (Beck et al., 1994; Giddens, 1990; Habermas, 1971).

As a consequence, science also has entered a much more uncertain time and age in which is asked to be agile and to be able to rapidly adept to changes in the real world. Nowotny et al. (2001) are politically not naïve, open to all kinds of interactions between science and society for all kinds of aims and goals, be it public, government and private. They conclude presenting a set of seventeen cultural, ethical, political and socioeconomic issues to be discussed in the agora where science and society meet. Nowotny et al., clearly go beyond the Legend, when they state that the 'epistemological core is empty', or 'there is no foundation' as it has been concluded by post-positivism (Chap. 12). Here they refer to the Legend and the lack of contribution analytical philosophy of science has had to the actual practice and methodologies of science in agreement with the discussion in Chap. 2. The value of research in the new way of doing research has, they argue in an outright normative stance, thus become dependent not on its 'eternal' abstract epistemic value, but on its reliability. Reliability and value in the epistemological meaning, but very much also when applied and tried out in action and set to work in the real. Here they refer specifically to Ziman's *Reliable Knowledge* (p157) (Ziman, 1978).

The thinking in Mode-2 differs very much from the 'solution' of Collins and Evans, two major scholars on the topic, who in 2017 wrote: *'In contemporary science and technology studies the predominant motif is to eliminate the division of powers between science and politics in order that science and technology become socially responsible. In contrast, our motif is to safeguard the division of powers so that science and technology can act independently of society'* (p7,8) (Collins & Evans, 2017) In a 'last defence', they argue that despite that we accept that no absolute value-free truth is produced by science, to regain its consequential loss of self-evident authority, science must be rescued by explaining science better and more honestly to society by those academics ('owls') that have that oversight of science and meta-science. Not by reflecting on the deficiencies of science, but through the moderation of these 'owls', they believe that society can be brought round to accepts the values of science. Then scientific experts can effectively play their role in the deliberations again. They discuss the work of relevant scholars, but the Mode-2 books discussed here are not mentioned. They do not trust the discourse in open debates with the public or within the scientific community, because of the interests and powers that are at play in non-ideal situations. They therefore do not take the, in my eyes, necessary reflexive next step to opening up science (p124–127).

Most of the critical scholars who's work I have discussed thus far, did blame the organization of science, its poor self-understanding, its flawed self-perception and some even courageously pointed at the reward system. They have, however, been much less bold with regards to proposing explicit ideas of how system change is to be made to the incentive and reward system to mitigate the observed problems. From the analyses presented in Chap. 3, it is clear why they did not discuss this highly sensitive issue. In contrast to most other scholars, Nowotny et al., do not at all duck at the difficult questions which relate to ideology and the self-understanding of science. In fifteen wonderful very confrontational pages which will still be disturbing for many scientists to read, they discuss the problems of The Legend (p50–65) (Nowotny et al., 2001). They have high hopes of Mode-2 research in which *'society speaks back to science'*. They at the same time realized that those who are still in the classical mode of science consider Mode-2 not 'real science', since they fear it obstructs 'real science' to be done. These critics will say it is not 'objective', disinterested and value free as science should be (Rouse, 1996; Douglas, 2009; Longino, 1990). Contextualization of science, in this classic vision, as we saw, is incompatible with the ideal and dream of 'objectivity' in the sense of the Legend. In line with post-positivist philosophy, Nowotny et al., argue that Mode-2 type knowledge production is done in a community of inquiry. Its claims are accepted by, and validated in social life and thus are intersubjective, reliable and socially robust. This is much more meaningful idea of 'objectivity'. Contextualization, that is starting with a problem and do the inquiry in that context or with that context in mind, is doing science the way of pragmatism. Despite that active scientists know that this is how science as a mature modern professional institution is being done, we saw that they are afraid to openly confess to their fallibility and limitations and are anxious of external influences and criticism. This, they believe, may hurt the image of purity and trustworthiness of their research. As Latour (1993) has concluded, this classical inward attitude and knee-reflex shows that science has not reflected and not made the full transformation, past positivism to real modernity. In our times in society it is recognized as 'scientism' which Habermas called 'halbierten Rationalismus', partial rationality, a science that operates from a positivistic frame in which it is insulated from social and cultural values.

Nowotny et al., are convinced of the opening up of science as the way forward to improve the impact of science. Science, the authors say, in our modern times should be reflexive and truly modern and thus have other worries: *'Today's scientists have to confront different, but analogous fears- their fear of the social world, with its imputed interests and ideological distortions, of cultural influences and of their own, subtle and not-so-subtle, accommodations to political and economic pressures....As public controversies proliferate, the trust of the public..has to be carefully nourished. If scientists would openly acknowledge these perceived threats, it might be possible to develop another model of knowledge production, in which knowledge becomes socially robust.'*

They argue that '*contextualisation has surreptitiously crept into what was once held the be the inner core of science whereas it has been embraced by more outward-oriented parts of science'* and argue that *...the actual practice of science ...might be set free to explore different contexts and perhaps to evolve in different directions...more as a comprehensive, socially embedded process' (p64, p65).*

Given the different ideologies, interests, fears and powers that are at play in the field of science, it is clear that this involves many actors and for some their most existential professional feelings. In short, this reorientation is not a small thing. Nowotny et al. describe the institutional changes toward the practices of Mode-2, for instance the alternative movements in the EU research area, where in Framework Programma (FP) 4 that ran from 1994–1998 a contextualized research programme was successfully launched which aims at targeted and problem-driven research programmes (EU, 2017). They use the term 'core' for the classical core of academia that is in Mode-1 and 'periphery' for the research and researchers that engage with problems and stakeholders from outside. They describe the tensions between them when governmental funding agencies are programming for problem-oriented research like in the EU FP 4 and 5. This is conceived by the core as 'undermining the peer review system' and the role of experts. Nowotny et al., correctly state that these directed programmes even are only 'Weakly Contextualized' as 'they were designed to solve yesterday's problems' and still operate in the classical linear mode of innovation. They discuss in depth the practice of Strong Contextualization, which starts with a policy agenda for research, prioritizing actual problems against each other as Kitcher has been describing in the ideal philosophical setting of well-ordered science.

The idea that Mode-2 was winning ground was not uncontested. In the boards and advisory committees of national and European Research Councils, they say, discussions, power struggles, between 'traditionalists' and 'modernists' in the complex field of science was pushing Mode-2 modernists out of the mainstream (high church) 'core', back into the 'periphery'. As the Mode-2 practice aims for robust reliable results and application, research, applied science and technology is being performed close to and many times with agents in the relevant societal contexts where urgent problems are being experienced and where it will present themselves to researchers. This type of research will have many different products, needs many forms of competences, skills and attitudes, demands different measures of quality control. Nowotny et al. wrote these observations on the practice of science in the present tense in the years just before the year 2000. Flyvberg, (2001) writing in the same years about 'how to make social science matter', points to this dynamics and to 'physics envy' as the wrong road for the social sciences. He makes for the social sciences a similar strong case for Mode-2 by revoking Aristotle's concept of *phronesis* which is a '*true state, reasoned, and capable of action with regards to things that are good or bad for man'* and *goes beyond* techne *and* episteme *since it is involved in social practice'* and he argues that '*attempts to reduce social science and theory to* <u>*episteme*</u> *(analytic) and or* <u>*techne*</u> *(technical knowledge) or to compre-hend them in those terms, are misguided' (p2).* These authors in the year 2000 cau-tiously, but optimistically, concluded that the practice of Mode-2 science and research -which is problem-driven, cross-disciplinary and pluriform in methods and approaches using modern and pre-modern humanistic ways to human understand-ing- had already reshaped part of science and research. However, we have seen in Chap. 3 that Mode-2 research in 2020 is still struggling with its image and standing in academia, despite the obvious problems of 'traditionalist' Mode-1 research.

5.6   Mode-2: Not the Highway of Academic Science

There thus have been strong local and even larger movements in academia and soci-ety in the 1960s and 1970s driving the case for social relevance. In addition, and derived from that until 2000, powerful and convincing academic analyses were pro-duced to show the urgency to optimally connect science with society and to remodel research and academia for that reason. As Nowotny (Nowotny et al., 2001) and others (Rip & van der Meulen, 1996) (Rip, 1994; Sarewitz, 2016) observed, in par-allel and in reaction to the inward looking well-organized academic community, in many countries a system of intermediary institutions and (semi-)government agen-cies had been established and was going to be even more firmly established to pro-gram science in Mode-2 style.(Whitley, 2000) Top down programming and research management *'seek to reconcile the upholding of standards of scientific quality with new demands that transcend them and need to be incorporated. The difficulty of setting priorities in funding in basic research highlights how the system is strug-gling to embrace a kind of social reflexivity- to which there is no alternative' (p47).* Nowotny et al., did realize that Mode-1 disciplinary science and scholarship, sci-ence for its own curiosity sake, aiming to add objective 'eternal' truths to the body of knowledge would not cease to exist. In their vision it is one of the consequences of the self-organizing capacity of science to manage *'the failure of scientific elites (Mode-1 elites) to accommodate to demands for accountability and priority setting and to accept additional criteria of judging the quality and relevance of scientific work'* (p47). They express their good hope for change in the very last lines: *'Just as 'publish or perish' is underpinned by certain rules of the game, to which scientists and their peers have agreed to adhere, so the opening up of science towards the agora presupposes and necessitates 'rules' of a game that partly still wait to be established.'...Not everyone will be able or willing to participate, and not anything goes-but the often feared 'contamination' of science through the social world should be turned around. Science can and will be enriched by taking in the social knowl-edge it needs in order to continue its stupendous efficiency in enlarging our under-standing of the world and of changing it. This time, the world is no longer mainly defined in terms of its 'natural' reality but includes the social realities that shape and are being shaped by science' (p262).*

Helga Nowotny: Reflecting on the Modes of Science

It was 7 November 2017 that I met Helga Nowotny at a half-day invitational workshop at the Robert Bosch Stiftung in Berlin. The debate was on '*Science and Science Policy: Is Knowledge Losing Power? Towards a More Resilient Science System for the 21st Century*'. Among the select group of participants were also Sir Mark Walport, Prof. Chief Executive Designate, UK Research and Innovation, United Kingdom, Sir Philip Campbell (Editor-in-Chief, Nature, London, United Kingdom), J-P Bourguignon (Director of the ERC), Dianne Hicks, Jack Stilgoe (UCL) and Tracey Brown (Sense about Science, UK). I was invited and explicitly instructed by the organisers to give the short opening statement of the workshop about Science in Transition in the 'provocative style'. I believe, judging from the report of the meeting*, that I managed to live up to these expectations. The reactions from the participants, given their positions in the field, were totally predictable. Walport and Bourguignon acted as if professionally offended, did not recognise the problem analysis at all. Campbell argued, as other publishers, publishers were not to blame. Stilgoe, Hicks and Brown joined me in adding their own critiques of science. Helga Nowotny, sitting at a corner of the table, did not appear to be shocked at all, but seemed rather slightly amused by the discussion. She quietly looked on. After the lunchbreak, Nowotny presented her reflections on the position and responsibility of science in society as to be found in here recent book *The Cunning of Uncertainty (Nowotny,* 2016*).*

I was familiar with Nowotny's work. We had discussed science as she gave a seminar at UMC Utrecht in our PhD course *This Thing Called Science* a year before, which was concluded with a small group dinner with Frank Huisman at the Faculty Club. Given the *Re-Thinking Science* book it puzzled me that she was also a founder of the ERC. Why in 2005 did she think we needed the ERC, a Mode-1 'high church' science build on the myth of Legend? That day in Berlin, I again wondered how to properly understand and interpret Helga Nowotny's work. According to her CV, on her website and interviews (Nowotny & Leroy, 2009) she was born in 1937, studied law in Vienna, and after that moved with her partner to New York and took to study sociology of science at Columbia in New York, studying with Robert Merton in the 1960s. Back in Austria in the 1970s, she researched political issues which involved scientific expertise, such as the debate about nuclear energy and got deep into STS ever since. She was affiliated with the Institute of Advanced Studies in Vienna, Faculty of Sociology, University of Bielefeld, École des hautes études en sciences sociales, Paris; Institute for Theory and Social Studies of Science, University of Vienna, professor at ETH Zurich. She was vice-president and until 2013 president of the European Research Council (ERC) of which she was one of its main founders. Reading her work, the books discussed in this chapter, but also her latest book *The Cunning of Uncertainty,* it is clear that Nowotny is a fine scholar and an exceptional

(continued)

expert on science, research and its interaction with society. She personally knows or has known the main scholars of her time, most of them discussed in these pages. She is completely aware of the conflicting interests, tensions, factions, politics and power struggles in academia. She was raised on the normative functionalism of Merton, knows the works of Habermas, Bourdieu, Foucault and Latour, but also has a broad overview of post-Merton, post-Popper philosophy and sociology of science. With this in mind, reading her work it is obvious that Nowotny at least in her writing evades the 'raw' politics of science. She does however not evade the problems of the Legend, she is leaving the myth of positivism behind and without explicitly mentioning Dewey, promotes the practice of American pragmatism. In my terminology, she moved from Mode-1 Legend to Mode-2 Pragmatism. Commentators on *The New Production of Knowledge* and *Re-Thinking Science* have correctly concluded that she and her co-authors seem scientifically and politically neutral with regards to Mode-1 and Mode-2 (Pestre, 2003). Mode-2, as the commentators argue, in their vision is thus open and vulnerable to the penetration by economic goals and the powers of the private sector. Nowotny et al., responded (H. Nowotny et al., 2003) adequately to these comments, taking all interests in scientific inquiry, economic and social into account, but staying out of issues of power and politics. The remark on the poor position they see for fundamental 'blue skies' research, explains the link with the ERC. Did they really believe that in academia Mode-2 would drastically displace Mode-1? I was inclined to read *Re-Thinking Science* as being a mix of descriptive and normative, that Mode-2 is a necessary complementation given the limitations of Mode-1. Nowotny even at the November 2016 small group dinner table in the Utrecht University Faculty Club could not be tempted to engage into too informal exchanges about the 'raw politics of science'. She reflects and presents the different options to us. To whom? I guess to the Deans and Vice Chancellors, boards of funding agencies who have to act in the real world once these academic insights have been presented.

*https://www.bosch-stiftung.de/en/project/berlin-science-debate/berlin-debate-2017

Despite these calls to contribute to the needs and urgent problems of society main-stream academia, science and scholarship in universities, Learned Societies, Royal Academies remained largely in Mode-1. Moreover, in academia the credit system with the typical metrics giving Mode-1 the highest esteem and the dominant academic career path was further embraced (Hicks et al., 2015; Wilsdon, 2016). This was not what the Mode-2 authors had hoped for. It was caused by a general development of government policies based on the idea that organized and programmed knowledge production is driving national economic and military competitiveness. This idea of 'the knowledge economy' started to fully play out in the late

1980s in the global economy for which the size and performance of national or regional (EU) science and technology systems were absolutely crucial. These national and in the EU international and regional science and innovations systems are however not to be thought of as 'an institutional set up somehow geared towards innovation. There is no inherent purpose of the overall system to work to some goal'(Rip & van der Meulen, 1996).

A multitude of non-synchronous interactions between the various bureaucratic organizations that acted as intermediate -sociologically designated as boundary-organizations on behalf of the government in relation to research organizations, universities and other public knowledge institutes (Whitley, 2000; Ziman, 1994). This was and is not a level playing field for researchers from the natural sciences, social sciences and the humanities and their subdisciplines. Some field thrived, others barely survived, waned or completely disappeared depending on internal academic ideas about autonomy, academic esteem and reputation; levels of proactive organization; external socioeconomic and all kinds of political developments. Probably the most significant of these 'fluctuations' after 1980 is the major decrease of investments mainly in physics and some natural sciences and the simultaneous enormous increase in biomedical and health research. This was directly related to the overall increasing expenditure on science and research, the end of the Cold War and amongst others the fight against cancer, increasing awareness of the effects of aging causing rapidly increasing health care expenditures (Stephan, 2012). Of note, for the same reasons, even within the field of biomedicine and health, but in fact in all fields, some researchers did benefit enormously compared to others from this increased public spending which depended mainly of what scientific advisory boards thought was excellent, held promise and should be funded. For instance, in biology and biomedicine from about 1970 on, due to major rapid breakthroughs in molecular biology, molecular genetics and protein chemistry, with its various more physical and chemical methods, this type of research in scientific advisory boards from many funding agencies became the norm for excellence. The molecular reductionist approach (*the molecular turn*) has ruled, from research on cancer, cardiovascular disease, paediatrics, infectious diseases, neurology to mental health and psychiatry. The role of the scientific committees at NIH, INSERM, CNRS, DFG, NWO, MRC, ERC, but also at Welcome Trust and other institutes and charities around the world is in this respect of interest (Miedema, 2012; Sarewitz & Guston, 2006). These institutes are at the boundaries were governments meets with universities and other public research institutes. It is their task to advance the societal missions and goals (the higher purpose) the government funder or charity has in view. On the other hand, to get the research properly done, it has to deal with the ideas and fashions and taste, about excellence, and internal politics of the different scientific fields (Lamont, 2010). This complex so-called 'principal- agent' problem has to be managed by esteemed scientists that are active in their fields and are appointed to the scientific committees. Complex it is, because the goals of funders and researchers in academia and public institutes are at best partially aligned. President Johnson, as mentioned above, in 1966 already expressed his doubts about the NIH. The MRC (Miedema, 2012) and most charities and also, as a national example the Dutch Cancer Society (KWF) for many years were 'hijacked' by the basic science approach of mainly molecular geneticists, disconnected from clinical care or public health. Writing in 2015 therefore about a new approach to funding, KWF courageously declared:

> *Our present approach to assessing grant requests is to focus on the scientific quality of the project or programme in question. From now on, we will place greater emphasis on the potential of the research in question to make a genuine contribution to our mission. The only way to obtain a clear picture of this is to examine each study in terms of its strategy for developing the results obtained into new treatments for patients. This calls for flexibility in the types of funding used: every effort must be made to ensure that the results genuinely reach patients. Throughout the entire research chain, from the laboratory to the patient, those working in the research field put forward research proposals, and we facilitate the flow of results. There is still a substantial focus on basic cancer research, as this is the source of new insights. Yet there is also scope for promising initiatives in the field of infrastructure, for example. https://www.kwf.nl/sites/default/files/2019-10/dcs-policy--vision-2015-2019.pdf*

The Bill and Melissa Gates Foundation must have realized this problem fully when the multi-billion Grand Challenges which started in the beginning of this century was evaluated. My laboratory was involved in a Mode-2 research project in which a large consortium, including researchers from Sub-Saharan Africa was addressing major problems on protective immunity to HIV/AIDS, malaria and TB with the goal of developing vaccines. Investments over ten years yielded many academic publications. The application, implementation and evaluation of the new knowledge in practice was a different game and appeared to be hard.

5.7   Re-visioning Science is Opening Up Science

Nowotny et al., in a chapter with the title Re-Visioning Science (Nowotny et al., 2001) briefly summarize the major issues with the following buzz words: realistic, reflexive, autonomy in localized forms, reliable knowledge, no universal objectivity, anticipate contexts. *'Co-evolution with society, demands a historically unprecedented openness on the part of science. Merely to add to the supposedly hard scientific core an additional outer layer consisting of 'softer' institutions, 'softer' norms and 'softer' behaviour on the part of scientists, all of which are designed to give greater weight to economic and social issues... cannot work. Science should attempt to reconstruct its image and authority ....: science is more heterogeneous, diverse, local and disunited* than the public and science itself realizes (p233). As I will discuss in Chap. 7, it took more than fifteen years to get broad institutional and international follow up of this call to the opening up of science.

References

Beck, U., Giddens, A., & Lash, S. (1994). *Reflexive modernization: Politics, tradition and aesthetics in the modern social order*. Polity Press.

Bourdieu, P. (1988). *Homo academicus*. Stanford University Press.

Bourdieu, P. (2004). *Science of science and reflexivity*. Polity.

Brown, M. B. (2004). The political philosophy of science policy. *Minerva, 42*(1), 77–95. https://doi.org/10.1023/B:MINE.0000017701.73799.42

Brown, M. B. (2009). *Science in democracy: expertise, institutions, and representation* (pp. 1 online resource (xvi, 354 pages)).

Brown, M. B. (2013). Philip Kitcher, science in a democratic society. *Minerva, 51*(3), 389–397. https://doi.org/10.1007/s11024-013-9233-y

Cartwright, N. (1999). *The dappled world: A study of the boundaries of science*. Cambridge University Press.

Collins, H., & Evans, R. (2017). *Why democracies need science*. Polity Press.

Dewey, J., & Rogers, M. L. (2012). *The public and its problems: An essay in political inquiry*. Pennsylvania State University Press.

Diggins, J. P. (1994). *The promise of pragmatism: Modernism and the crisis of knowledge and authority*. University of Chicago Press.

Dijstelbloem, H. (2014a). Missing in action: Inclusion and exclusion in the first days of AIDS in the Netherlands. *Sociology of Health & Illness, 36*(8), 1156–1170. https://doi.org/10.1111/1467-9566.12159

Dijstelbloem, H. (2014b). Science in a not so well-ordered society. A pragmatic critique of procedural political theories of science and democracy. *Krisis, 1*, 39.

Dijstelbloem, H., & Hagendijk, R. (Eds.). (2011). *Onzekerheid troef: het betwiste gezag van de wetenschap, Amsterdam: Van Gennep*. Van Gennep.

Douglas, H. E. (2009). *Science, policy, and the value-free ideal*. University of Pittsburgh Press.

Epstein, S. (1996). *Impure science: AIDS, activism, and the politics of knowledge*. University of California Press.

Epstein, S. (2007). *Inclusion: The politics of difference in medical research*. University of Chicago Press.

EU. (2017). EU frame work programmes for research and innovation evolution and key data from FP1 to Horizon 2020 in view of FP9. Retrieved from.

Felt, U., & Fochler, M. (2010). Machineries for making publics: Inscribing and de-scribing publics in public engagement. *Minerva, 48*(3), 219–238. https://doi.org/10.1007/s11024-010-9155-x

Flyvbjerg, B. (2001). *Making social science matter: Why social inquiry fails and how it can count again*. Cambridge University Press.

Gibbons, M., Limoges, C., Nowotny, H., Schwartzman, S., Scott, P., & Trow, M. (1994). *The new production of knowledge: The dynamics of science and research in contemporary societies*. SAGE Publications.

Giddens, A. (1990). *The consequences of modernity*. Polity in association with Blackwell.

Habermas, J. (1968). *Technik und Wissenshaft als 'ideologie'*. Suhrkamp verlag.

Habermas, J. (1970). *Toward a rational society*. Heinemann Educational Books.

Habermas, J. (1971). *Knowledge and human interests*. Beacon Press.

Hicks, D., Wouters, P., Waltman, L., de Rijcke, S., & Rafols, I. (2015). Bibliometrics: The Leiden Manifesto for research metrics. *Nature, 520*(7548), 429–431. https://doi.org/10.1038/520429a

Irwin, A., & Wynne, B. (1996). *Misunderstanding science?: The public reconstruction of science and technology*. Cambridge University Press.

Jasanoff, S. (2012). *Science and public reason*. Routledge.

Kitcher, P. (1993). *The advancement of science: Science without legend, objectivity without illusions*. Oxford University Press.

Kitcher, P. (2001). *Science, truth, and democracy*. Oxford University Press.

Kitcher, P. (2011). *Science in a democratic society*. Prometheus Books.

Lamont, M. (2010). *How professors think: Inside the curious world of academic judgment.*

Latour, B. (1988). *The pasteurization of France.* Harvard University Press.

Latour, B. (1993). *We have never been modern.* Harvard University Press.

Longino, H. E. (1990). *Science as social knowledge: Values and objectivity in scientific inquiry.* Princeton University Press.

Marres, N. (2007). The issues deserve more credit: Pragmatist contributions to the study of public involvement in controversy. *Social Studies of Science, 37.* https://doi.org/10.1177/0306312706077367

Miedema, F. (2012). *Science 3.0. Real science real knowledge.* Amsterdam University Press.

Miller, J. (1994). *"Democracy is in the streets": From Port Huron to the siege of Chicago: With a new preface by the author.* Harvard University Press.

Nowotny, H. (2016). *The cunning of uncertainty.* Polity.

Nowotny, H., & Leroy, P. (2009). Helga Nowotny: An itinerary between sociology of knowledge and public debate. *Natures Sciences Sociétés, 17*(1), 57–64. Retrieved from https://doi.org/10.1051/nss/2009010.

Nowotny, H., Scott, P., & Gibbons, M. (2003). Introduction: 'Mode 2' revisited: The new production of knowledge. *Minerva, 41*(3), 179–194. https://doi.org/10.1023/A:1025505528250

Nowotny, H., Scott, P., & Gibbons, M. (2001). *Re-thinking science: Knowledge and the public in an age of uncertainty.* Polity/Published in the USA by Blackwell.

Oudshoorn, N., & Pinch, T. (2003). *How users matter: The co-construction of users and technologies.* MIT Press.

Pestre, D. (2003). Regimes of knowledge production in society: Towards a more political and social Reading. *Minerva, 41*(3), 245–261. https://doi.org/10.1023/A:1025553311412

Price, D. J. d. S. (1963). *Little science, big science.* Columbia University Press.

Rawls, J. (1999). *A theory of justice* (Rev ed.). Oxford University Press.

Rip, A. (1994). The republic of science in the 1990s. *Higher Education, 28*(1), 3–23. https://doi.org/10.1007/BF01383569

Rip, A., & van der Meulen, B. J. R. (1996). The post-modern research system. *Science and Public Policy, 23*(6), 343–352. https://doi.org/10.1093/spp/23.6.343

Rouse, J. (1987). *Knowledge and power: Toward a political philosophy of science.* Cornell University Press.

Rouse, J. (1996). *Engaging science: how to understand its practices philosophically* (pp. 1 online resource (ix, 282 pages)). Retrieved from http://purl.oclc.org/DLF/benchrepro0212

JSTOR. Restricted to UCSD IP addresses http://www.jstor.org/stable/10.7591/j.ctv75d2nk.

Sarewitz, D. R. (2016). Saving science. *The New Atlantis, 49,* 4–40.

Sarewitz, D. R., & Guston, D. H. (2006). *Shaping science and technology policy the next generation of research.* University of Wisconsin Press.

Spaapen, J., H.Dijstelbloem F. Wamelink. (2007). *Evaluating research in context. A method for comprehensive research assessment.*

Stephan, P. E. (2012). *How economics shapes science.* Harvard University Press.

Toulmin, S. (1977). From form to function: Philosophy and history of science in the 1950s and now. *Daedalus, 106*(3), 143–162.

Toulmin, S. (1990). *Cosmopolis: The hidden agenda of modernity.* Free Press.

Whitley, R. (2000). *The Intellectual and Social Organization of the Sciences,* vol 11.

Wilsdon, J. (2016). *The metric tide: The independent review of the role of metrics in research assessment & management.* SAGE.

Wilsdon, J., Stilgoe, J., & Wynne, B. (2005). *The public value of science: Or how to ensure that science really matters.* Demos.

Wilsdon, J., & Willis, R. (2004). *See-through science: Why public engagement needs to move upstream.* Demos.

Ziman, J. M. (1978). *Reliable knowledge: An exploration of the grounds for belief in science.* Cambridge University Press.

Ziman, J. M. (1994). *Prometheus bound: Science in a dynamic steady state.* Cambridge University Press.

Chapter 6
Science in Transition Reduced to Practice

Abstract In the true spirit of Dewey and pragmatism, knowledge, insights and experience have to be translated into interventions and actions. Only when knowledge is 'reduced to practice' its social robustness and value will be determined. In light of the conclusions of the previous Chapter, to be able to have more impact and to hold up our promise to society we have to reflect who our science is organized and how it could be improved. From these reflections, several interventions in the practice of research have been proposed. When we, the Science in Transition Team, started to make public our critical accounts of the practice of science, I was 'friendly advised' by influential older scientists to first clean up the mess in my own institution, instead of pointing to others and to the system. As a matter of fact, that is what we have been doing at University Medical Centre Utrecht (UMC Utrecht) since 2009. In this chapter I present a brief outline of our actions 'on the ground' in UMC Utrecht and some early actions to promote these activities abroad.

UMC Utrecht is a large academic medical centre, to give an impression, key figures of 2009 that I typically used to show in my introductory talks about UMC Utrecht, are depicted below. I am happy to see that the slide did not show JIF's and numbers of citations, the position on the Shanghai Ranking and amount of grant money won. At that time, some 1200 researchers were working on a PhD thesis project, which saw an enormous increase over the 15 years before.

© The Author(s) 2022
F. Miedema, *Open Science: the Very Idea*,
https://doi.org/10.1007/978-94-024-2115-6_6

University Medical Center Utrecht
Research, education and care

- The UMC Utrecht was founded in 2000 through the merger of the
  Academic Hospital, Wilhelmina Children's Hospital (WKZ) and the Medical
  Faculty of Utrecht University

- *11.000 employees,*
- *1000 beds,*
- *3500 students,*
- *2200 scientific papers*
- *200 PhD thesis defenses*
- *2,340 births*
- *634 deceased*
- *41,400 hours of surgery*
- *1,620,928 website visitors*
- *820,000 meals*
- *2,210 tons of waste*

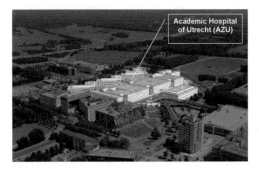

6.1   The Matrix

In September 2008 in the USA the Lehman Brothers Bank was not bailed out and
had fallen. Because of these and other ominous signs in the months before, we realized that the financial crisis was imminent and visible and was going to hit major
banks and financial instituters in Europe as well. We anticipated a serious collapse
of the economy in Europe but also in our country when we started the first of January
2009 in a new composition of the Board of UMC Utrecht. The chairman, Professor
Jan Kimpen was a paediatrician who had, before joining the board, been chairman
of the Wilhelmina Children's Hospital, a division of UMC Utrecht. The third board
member, Herman Bol, came from the financial sector. I had left Sanquin in 2004 to
chair the department of immunology and of the Division Laboratory and Pharmacy
since 2005. After a couple of months, we started to work on a new five-year strategy.
We had held discussions with our regional partners, the partners in university and
corporate partners and patient-advocacy groups. We, in fact our our staff, evaluated
the two strategies of the past decade and looked at potentially interesting examples
of institutional research strategies abroad. The conclusions were quite interesting
and refreshing. Our UMC had since 2000 been organized as a collection of divisions, small hospitals each based on a set of related medical disciplines like internal
medicine, surgery, paediatrics, neurology and psychiatry, gynaecology, cardiology
and pulmonology. A number of divisions were about enabling methods and technologies, as laboratory sciences, epidemiology, medical imaging, radiology,

molecular biology and clinical genetics. The divisions were very well organised, performed according to finance- and production-related key performance indicators (KPI's) very well. The institute had because of that an excellent financial position. The divisions had their own overall strategies and goals which were discussed yearly with the board. For research, there had been a top-down formulated five-year strategy, which the organization had experienced as very nice but abstract, not including very concrete milestones and goals. As division management was held accountable for staying with in their budgets, the incentives for entrepreneurship were low and collaborations over 'the borders' of divisions was problematic. There was a wide gap between basic pre-clinical research, most of it done in a semi-separate building and the more clinically oriented research in the hospital. The people in interviews complained that this was inefficient for research and innovation of care but also for daily delivery of clinical care. This down-side to the governance model had been consciously considered against its advantages in 2000 by our predecessors in the Board, but as the institute came from an instable financial situation in 1998, sound financial results were the first priority, and it was hoped that the organizational issues could be mitigated by wise leadership in the divisions. In our opinion, based on the evaluation of the past ten years, this appeared to have become increasingly problematic and required an intervention to facilitate and incentivize necessary collaborations between the divisions, both in clinical care as in research and innovation.

In light of this, after much deliberation, we decided to aim for a maximum of six large strategic research programmes that should be goal-oriented and connect relevant classical disciplines and divisions. The programmes should by definition thus be multidisciplinary, bring the more fundamental, pre-clinical work in the context of the relevant clinical departments or extramural domains of prevention and public health. We anticipated that the programmes would be quite large but still should be focussed on a small number of concrete short-term and long-term public health or clinical targets. These programmes, we emphasized, should truly aim for impact in science and society. A small group of professors drafted 'terms of reference' to provide guidance to the writing of proposals and broadly also defined criteria for quality and feasibility once choices had to be made. Based on this groundwork, we invited our professionals to present ideas for strategic programmes.

In a one-day session early in 2010, with forty senior colleagues in the room, we democratically picked 22 of the best proposals from over sixty proposals that had been submitted. They were merged into six major disease-oriented programmes that each covered for their domain the whole spectrum of basic, applied and clinical research from the disciplines that involved. For example, in Personal Cancer Care researchers from epidemiology, medical oncology, molecular genetics, surgery, radiology but also representatives of patient advocacy groups and other stakeholders participate. Eventually, it was realized that we needed rehabilitation sciences and bioethics in such programmes as well. Because of the interactive loop, beyond the linear model, it was decided in the board that we called the new strategy *UMCU 3.0.*

6.2 The Innovation Loop

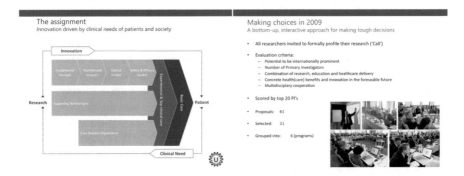

The assignment
Innovation driven by clinical needs of patients and society

Making choices in 2009
A bottom-up, interactive approach for making tough decisions

- All researchers invited to formally profile their research ('Call')
- Evaluation criteria:
  - Potential to be internationally prominent
  - Number of Primary Investigators
  - Combination of research, education and healthcare delivery
  - Concrete health(care) benefits and innovation in the foreseeable future
  - Multidisciplinary cooperation
- Scored by top 20 PI's
- Proposals: 61
- Selected: 21
- Grouped into: 6 (programs)

It is a bit beyond the scope of this book, but obviously most critical for the change-management we had engaged in, were management and administrative problems that come with building such multidisciplinary programs in a matrix organization with ten divisions that are disciplinary with respect to science and medicine. Building this type of programmes required collaboration and discussion across the classical boundaries of basic-applied, preclinical – clinical and between the different clinical disciplines that were organized in divisions with strong classical structures. Clinical disciplines are far more distinct that disciplines and fields of research. They link up to years of professional medical training and clinical work with severe and often distinct patterns of socialization for their professionals. In the case of oncology, professionals from internal medicine, pathology, surgery, radiotherapy, geriatrics and rehabilitation medicine need to seamlessly work together to achieve optimal patient care. As I discussed with regards to research, also here professional hierarchies, in which the professionals are socialized, covertly or explicitly are at play. In an academic (university) medical centre the 'field' of research is intertwined with 'the field' of the medical profession each with their own power struggles and stratification as described in Chap. 3 for research. With Bourdieu's *Distinction* in the back of my mind, I often realized that surgeons and internal medicine doctors are very different people indeed (Bourdieu, 2010). The complexity of this dual world of science and medicine should not be underestimated.

Because of this, successfully building consortia to form strategic programmes, requires real leadership and brinkmanship from many senior professionals. Negotiations between program and divisions about choices to be made regarding research topics and clinical work, investments and joint decisions about human resource management, hiring and promotion of personnel, were complex. These issues of '*alignment to the higher purpose*'* are classical and abundantly discussed in the literature on innovation, R&D and research management in research intensive industries (* I borrowed 'the higher purpose' from Manfred Kets de Vries who has advised us on management issues in 2015). Even in institutions and corporations, like Philips or pharmaceutical companies were problem- and product-oriented research is normal, and despite a much higher corporate identity and shared value with the overall corporate goals, these interactions pose managerial challenges. It took literally years for, us, the institute to get used to the new organizational scheme. During my time as director of research in Sanquin, given the mission of Sanquin, our research was for a large part to be directed at the development of products and services related to development and safety of blood products. In those days I read the literature about managing top professionals and now and then in UMC Utrecht I went back to some of those books such as Maister's *True Professionalism* and *Third Generation R&D* by Roussel, Saad and Erickson (Maister, 1997; Roussel, Saad, & Erickson, 1991). Interestingly, Mirko Noordegraaf and Paul Boselie, colleagues from the Utrecht School of Governance in 2012 showed interest to study our management intervention as a real-live case in their long-term research programme on public management. The key question in that program is how public organizations and private organizations with a public task deal with current social issues, how they shape their public responsibilities and deliver public value. Also though Noordegraaf's contact in an EU project I was invited to present our case in Bologna in April 2012 to thousand (!) representatives of hospital management in the Italian Region of Emilia Romagna. Later, in September 2014 I did my talk at a meeting with the Karolinska and Stockholm hospital system. Noordegraaf and Boselie joined forces with Margriet Schneider to establish the Utrecht University Focus Area Professional Performance in 2015. Margriet Schneider then was chair of the Division of Internal Medicine and later that year became Chair of the Board of UMC Utrecht.

The next five-year strategy, was initiated in 2014 when Mirjam van Velthuizen-Lormans, who had a long career already in UMC Utrecht before, became Board member. With this strategy that started January 2015 we took it to the next level. The 'innovation loop' was shown to be totally interactive engaging with regional, national and international academic and non-academic partners outside the walls of the institute. Therefore, the strategy was appropriately called 'Connecting U'. When I proudly presented the strategy at a 2014 Christmas party to our retired colleagues, a lady in the front loudly remarked that she thought it was a nice name, but perhaps better suited for an Utrecht public transport company. I agreed of course with a big smile, but politely retorted that we felt it was also very appropriate, and nice for an academic hospital serving and connecting with the greater Utrecht region.

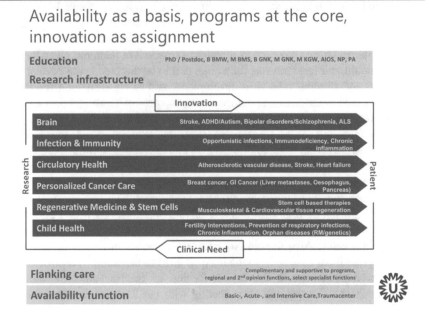

Availability as a basis, programs at the core, innovation as assignment

Education   PhD / Postdoc, B BMW, M BMS, B GNK, M GNK, M KGW, AIOS, NP, PA

Research infrastructure

Innovation

Brain	Stroke, ADHD/Autism, Bipolar disorders/Schizophrenia, ALS
Infection & Immunity	Opportunistic infections, Immunodeficiency, Chronic inflammation
Circulatory Health	Atherosclerotic vascular disease, Stroke, Heart failure
Personalized Cancer Care	Breast cancer, GI Cancer (Liver metastases, Oesophagus, Pancreas)
Regenerative Medicine & Stem Cells	Stem cell based therapies Musculoskeletal & Cardiovascular tissue regeneration
Child Health	Fertility Interventions, Prevention of respiratory infections, Chronic Inflammation, Orphan diseases (RM/genetics)

Clinical Need

Flanking care   Complimentary and supportive to programs, regional and 2nd opinion functions, select specialist functions

Availability function   Basic-, Acute-, and Intensive Care, Traumacenter

6.3 How Do You Want to Be Judged?

There had been in our institute for over fifteen years already a strong focus on research. In line with what at these days was pretty normal, the more fundamental pre-clinical science was regarded the best, as was measured by the JIF of the venue of publication. Publication output, citations, JIF and top 10% of journals of the field and in addition high profile personal grants and the amount of grant money that was brought in were used to score the research performance of divisions. This type of metrics was every three-year period used to determine the number of professors each of the divisions was entitled to have. As the total number of professorships was capped, this was a zero-sum game where every three years some divisions lost, some gained. Fortunately, only a limited fraction of intramural money was allocated to the divisions based on these indicators, in addition to monetary rewards for the number of awarded PhD's. As we have seen in Chap. 3, this was since the years 2000 common practice in the Dutch and the wider European and international research landscape. UMC Utrecht in that respect was not at all atypical. In this strategy we did very well in publications, numbers of PhD's and grant money that was brought in. With our new strategy a year or two underway, however, we after some time had to admit that this incentive and reward system was not aligned with the different forms of science and academic output in the six multidisciplinary programs. In fact, we also realized that our research evaluation system did not acknowledge top professionals and clinicians engaged in more practical patient-driven research where journal impact factors are lower and no prestigious personal career grants are to be won. The more your research was to the left of the innovation loop, the better your chances were for high JIF publications and thus academic promotion. Of course, there were exceptions. When regarding clinicians who performed extremely well and were scarce because they did and thought surgery at stellar levels, after fierce debates in academic appointment advisory committees, their publication lists and grants won were regarded less important compared to their professional academic performance and impact.

The Higher Purpose

In 2014, the need to change the system of research evaluation forced itself upon us in UMC Utrecht. This was a couple of years after we had made the change in the organization of our research environment. For us this was quite logical, conceptually and in time but, to be honest, it had not been planned in 2010. This struck me again in January 2017, at a Washington DC bookstore-restaurant having a breakfast meeting with Paul Wouters and Dan Sarewitz. The three of us attended a special one-day meeting on incentive and rewards organized by Metrics Stanford. Sarewitz made it quite clear, he was not much into the problem of metrics, but had been thinking for decades about the organization of science and how to effectively change it. Sarewitz is well-known from his critical well-informed pieces about the science system in *Nature* and his book, *Frontiers of Illusion*, his book chapters and his opinionated excellent long read *'Saving Science'* (Sarewitz, 1996, 2016). His work has focused on the politics of science and how all kinds of forces and powers keep science from living up to the promise to optimally contribute to society and the good life. He is highly critical of those who are pursuing intellectual interests of 'blues skies' research with reference to the linear model of innovation and value-free inquiry. It is an endless frontier, but in his analysis with a lot of illusion indeed.

Sarewitz, asked me why and how we had been able to agree on and then implement a new system of research evaluation at UMC Utrecht. I told him the story of our intervention in UMCU 3.0 and that it thus was a logical consequence of our strategy. It was the diversity of goals and academic roles defined in the six strategic programs, that after a couple of years forced us to implement a research evaluation system that matched with these goals and with the *'higher purpose'* of UMC Utrecht. We had assigned this task to a group of midcareer young researchers and clinical professionals chaired by Marieke Schuurmans, professor of Nursing Science and secretarially supported by Rinze Benedictus who was by then already quite an expert on Incentives and Rewards. We invited them in August 2015 to start on the question 'How do you want to be judged and evaluated?' After six months they presented a more inclusive and less metrics-driven evaluation protocol. The result was a very open and generic scheme which allowed to honour the pluriform excellences related to the diversity of academic roles in the system. Not only papers published, or funding obtained had to be considered, but also

(continued)

results being used and applied closer to users by peers or by users and stake-holders themselves. Think of application in the clinic, in medical products and technical appliances via private partners, in a treatment advise by the Health Council, in the organisation of health care in the region, or in policy making of any kind. In addition, a lot of emphasis was on the *ex-ante*, or 'how' the research was organised in order to enhance on beforehand its potential impact. For instance, we asked, if there was early engagement of stakeholders. The scheme and its implementation were not uncontested. Some warned that 'it would come at a cost to the quality of our research, was to hurt basic science and the reputation of our institute. 'It very much depends on how, and who defines 'research quality', was my response. Of course, although we all believed we were moving in the right direction, I very well understood the issue. The risk of a first-mover disadvantage posed a serious and realistic worry in those days when DORA was barely known and there was a global addiction amongst academics and university administrators to JIF, h-indexes and the Shanghai Ranking. Even in 2020, when a lot has happened regarding Incentives and Rewards, nationally and internationally, understandably this is the worry still most frequently aired by young research professionals.

The worry about basic science, as we have seen in the previous chapters is of all times. Here I refer to Stoke's *'Pasteur's Quadrant* where the concept of 'user-inspired basic science' is explained as the kind of research most researchers in basic science do (Stokes, 1997).

UMC Utrecht: Inclusive set of generic indicators for research quality and impact

Structure	Leadership & culture
	Collaborations with stakeholders
	Continuity and infrastructure
Process	Setting research priorities
	Posing the right questions
	Incorporation of next steps
	Design, conduct, analysis
	Regulation and management (OA, FAIR data sharing)
Outcomes	Research products for peers
	Research products for societal groups
	Use of research products by peers
	Use of research products by societal groups
	Marks of recognition from peers
	Marks of recognition from societal groups

 Universiteit Utrecht

User-inspired basic science takes on problems in the context of a larger problem in a given practice and investigates 'blind spots' and missing links in knowledge and understanding in that particular field. As we have seen, basic science has a higher standing than applied science, even with the public, and this still feels like a problem for the investigator. In a typical early evening show that until recently ran on Dutch TV and was famous for a host with boundless admiration of scientists, we often see the invited scientist first explain how terribly fundamental the work is, to demonstrate its scientific quality, in order to then proudly explain how it can be used to solve a clinical or social need. Even our recent Nobelist, the synthetic organic chemist Ben Feringa, who started his career at Shell Research, did not escape this knee-jerk reflex when in 2016 he explained in the Dutch evening news his price-winning work as totally 'blue skies', but a moment later proudly explained that his molecular motors once may be used to direct medicines to the right spot in the body of patients among other applications in practice.

How to Make the Right Choices
One day, at my job in the department of immunology of UMC Utrecht, I got a phone call from my sister that my brother, who was more than ten years my senior, had suffered a very serious stroke and was in hospital. I went to see him at the hospital, near where he lived. He was in very bad shape. It was a devasting sight. He was paralysed on his left side, but the most terrible thing was that he could not speak and probably had serious cognitive problems. He was moved to a well-known rehabilitation centre in the heart of Amsterdam. During visits we sat in a common room, with a view of the Vondelpark. His ability to move the left leg and arm returned pretty quickly. His speech did not return and communicating with him during visits and ever thereafter was very difficult, which frustrated him enormously. Looking around at the facility, its ambiance, shocked also by the sight of also relatively young patients and their visitors, I was reading the information leaflets about the rehabilitation therapy my brother was receiving. I could not help myself to think of the enormous investments made over the years in research on the pathogenesis of stroke, involving numerous PhD positions, sophisticated animal models, laboratory equipment and large expensive devices and the most innovating molecular and imaging technologies. As the fast majority of patients survive stroke but badly need medical rehabilitation for recovery of speech and mobility and cognitive recovery, the low academic priority and very modest investments in innovation in research and development of rehabilitation and mobility research, I realised, were a disgrace.

After a few years I became the dean and I was confronted with this problem in my own UMC Utrecht and later realised it had been noted at that time by the national Health Council. Because of the reward system, its metrics and definitions of excellence, rehabilitation sciences were suffering. Typical career advice to young MD's therefore was: 'go for a PhD on a topic of 'hard science'

(continued)

such as molecular pathogenesis. It has more esteem, gives better papers and a better CV than to work on applied problems of mobility and rehabilitation'. Be sure, such problems caused by 'the system' is nobody's fault. People, even highly educated people 'read the system', behave according to the system and adept strategically to seek possibilities of advantage for themselves and their set up. I could fill many pages with similar problems of agenda setting being distorted by the incentive and reward system. Molecular cancer biology versus research on living with adverse effects of chemo, total immune ablation, radio-therapy and a bone marrow transplantation. The tumour is hopefully gone, but the patient is still there struggling with her poor quality of life.

6.4   The Call for Health from Society

At that time, a general resistance was rapidly rising against the dominant idea that even in the public sphere, literally all public services, should be left to private par-ties and in our case the market of health care. The classical economist's idea was that 'automatically' this competition would result in more efficient and cost-effective services, compared to the situation in which non-profit semi-government organizations offered these services. This neoliberalism (and globalization), with its focus on the mechanism of the (international) corporate markets and competition steered by shareholder value or the principles of New Public Management, however, did not apply to schools and higher education, but also not to health care. Apparently, these services are not typical consumer products, but more of the type of common goods essential for the quality of social life and of the public sphere in civil society and democracy and must be regulated and provided through government. In the Netherlands but also in the wider EU in a similar vein, politicians both liberal-conservatives and social-democrats realized the down sides that the politics of the Third Way have had. In in our country since 1994 this is designated as 'Paarse Politiek'. Science in Transition did not put all of the blame for problems in aca-demia on the politicians and government, that was thought to be too easy. We real-ized and showed, as discussed in Chap. 3, that academia and academics, but also administrators in university and other academic and funding institutions had quite willing adjusted their strategies and practices to these neoliberal policies.

The Netherlands Health Council in response to a request of the Minister of Health, Welfare and Sport, in October 2016 published an advice on how to improve the impact on the Dutch health care system by the research done in the eight University Medical Centres. In the letter of request to the Health Council, to our pleasant surprise, the minister cited the analysis of Science in Transition on 'how metrics shapes science', and would that not be a problem? The committee installed by the Health Council that produced the above-mentioned advice was clear: to a great extend research is not driven by the needs of public health, of the care or cure system, but is too much focussed on research driven by parameters of esteem, clearly related to the metrics used in academia. The report specifically pointed out the fields with high societal and clinical relevance that got too little research attention and investment in the current system. These included mostly public health and prevention and research to improve the health care system focused on problems in the region around UMC's and national issues. The mismatch between investments in biomedical research and disease burden at patient and societal level has over the years been regularly reported in *Lancet* or *BMJ*. The novelty was that it was causally linked to the perverse effects of the incentive and reward system. The Council, citing the relevant international literature, understood that researchers, make strategic choices in which field and on what topic to do research. That this was increasingly based on the chances of building a resume mainly with particular journal articles to get credit and esteem from peers, required for the next round of grants. The current JIF dominated metrics game, the committee concluded, steers researchers away from the fields where they are closer to patients in the wards and away from citizens in the region. Is seemed as if the idea was, in my words, *'the further away from the patient, the cleverer you are'*. The Boards of UMC's were not all that amused by the Health Council's advice, and in a knee-jerk reflex which made it to the frontpage of a national newspaper it was rebutted that the Council did not show respect for 'the beautiful basic research with high international visibility that is being done in UMC's which is the basis for excellence in Dutch health research'. Initially the usual evasive and defensive voices played the Council's critique down, saying that there was not at all a problem with regional collaboration. After some months of discussion though, it was realized that the Health Council and the Minister were to be taken very seriously. It was a problem for patients, the public and society at large and thus action required from the UMC's, since most of their research was paid for by tax money. With professor Albert Scherpbier, the Dean of UMC Maastricht, in the lead we got a group of national experts together to compose an action plan for the UMC's, for which we consulted virtually every stakeholder in society. As a result, a bold plan was designed in response to the Health Councils advice.

https://www.nfu.nl/img/pdf/19.5200_Research_and_Innovation_with_and_for_the_healthy_region.pdf

The plan basically was to adjust the research agenda to better respond to societal needs with regard to public health, prevention and clinical care. A clear shift to more regional and national societal impact was one of the major aims. The UMC's committed themselves to setting up a regional network around each UMC to deliberate on the most urgent problems and how to work together, through research and social action, to improve cure, care, health and welfare in the region. This transition was not going to be easy, as was realized by the Council and the UMC's, this would not happen without explicitly changing the incentive and rewards system of research and researchers in the UMCs. Proper incentive and rewards are required to acknowledge the diversity in research excellence of researchers in for instance public health doing quantitative social research and work on lifestyle, nutrition, and quality of life in mental disease. In our own backyard in the larger Utrecht region, in line with this advice, we had already invested in setting up this type of regional collaborations for the treatment of high complex rare cancer types with the four hospitals https://www.umcutrecht.nl/nl/ziekenhuis/regionaal-academisch-kankercentrum-utrechtraku,and had initiated a round table *Gezond Utrecht* that brought together all health care providers in the region. https://bestuurstafelgezondutrecht.nl

During the COVID-19 crisis in the spring of 2020, this type of non-competitive collaborations were top-down enforced and became highly visible when health care, and in particular clinical ward and ICU capacities, personal protective equipment and the testing for COVID-19 had to be nationally and regionally organized. With respect to the need to improve collaboration and dismiss market-type competitions between cure, care and health providers, it became clear that COVID-19 had the most devasting effect on the elderly that were not in hospital, but in dedicated care homes. In the first two months of the pandemic internationally the focus was on the ICU and hospital care and cure but medical care in care homes did get little attention. We thus were confronted with the hierarchy in the medical profession between cure and care but also of insufficient 'scientific' appreciation of research and innovation potential of fields like geriatrics, rehabilitation and preventive medicine, despite its immense social impact in our ageing populations, not only times of Corona. We now hear a serious call to rethink this policy, will it last after COVID-19 is under control? Do we, in 'the cold phase', then still want to invest and pay more to have better and reliable availability of medicines and be better prepared for the 'hot phase' of the next pandemic?

6.5 Science in Transition Abroad

We reassured our public that our initiative and most of our agenda was part of a larger emerging international movement. In all our talks we therefore took great pains to point to some of the high-profile initiatives already ongoing abroad in biomedical research. Most of them as discussed in Chap. 3 were mainly focused on quality with respect to design, clinical impact and reporting of research and not so much about change of the system. With two of those initiatives we got connected in 2015. I think this was good for visibility and crucial for our credibility, nationally as well as internationally. Apparently, as I had learned in the field of AIDS research 30 years before, also in the field of meta-science, although its funding was at least three orders of magnitude smaller and less defined as a research field, one had to spot the 'right people' and their international network to connect with, in order to enhance impact. I will pay due credits to two very different men.

6.6 EQUATOR, Meeting a True Pioneer: Doug Altman (1948–2018)

February 19, 2014 a small-format symposium was held at UMC Utrecht under the title Science in Transition. I was asked by the organiser, Carl Moons, to say a few words of warm welcome as an introduction to Doug Altman, as the Dean is expected to do. It was a month after the publication of the high-profile series of papers in *The Lancet* of which Doug Altman was a major initiator and author. (discussed in Chap. 3) Altman was one of the co-founders of *The EQUATOR health research reliability network* and has written major papers since the beginning of the century on quality issues in biomedical research and its reporting. Some of his early papers in BMJ are classics that are still widely read. In the midst of the launch of Science in Transition, knowing his strong position about science, I did not do my courteous Dean's intro, but a strong pitch 'how science went wrong and what should be done about it'. Altman, a bit surprised at first, but then feeling free to speak up, very British but passionately gave his talk. The next year, as he obtained an Honorary Doctorate from Utrecht University, we met again. Doug was a very nice, soft-spoken scientist who was really worried about the quality of science and not fond of a lot of attention and being in the spotlights. I still think though, he was truly pleased with 'the honour bestowed upon him' by our University.

I believe as a result of these informative interactions during dinners before and after the University Dies Natalis of March 2015, Dough invited me to speak at the Reward/Equator Conference in Edinburgh, September 2015. This was a meeting organized by the group of authors of the Lancet papers of January 2014. (P. Glasziou, 2014; Paul Glasziou et al., 2014; Macleod et al., 2014) It were mostly biostatisticians and methodologists, most of whom were present. In fact, many major players from all over the globe, who actively worked to improve biomedical science, at journals, funders and universities were present. I was the only one that day arguing for a systems approach to the problem from the 'Dean's Perspective'. My call was that we needed to break free from that perverse credit cycle, but in order to do so we needed to engage the people how have power in the system: university administrators, deans, board members of the Royal Societies, patient advocates, charities and government funders. Unfortunately, too few of them were to be found in the audience that day.

6.7 METRICS, the Relentless John Ioannidis

From the speaker's podium he could not be overlooked, seated attentively in the front row, dressed in his habitual spotless white summer costume, often complete with a bright red tie. John Ioannidis (1965) is C.F. Rehnborg Chair in Disease Prevention, Professor of Medicine, of Health Research and Policy, of Biomedical Data Science, and of Statistics; co-Director, Meta-Research Innovation Centre at Stanford. Ioannidis is a Greek-American physician-scientist and writer who has made contributions to evidence-based medicine, epidemiology, and clinical research. Ioannidis nowadays is well-known for his studies of scientific research itself, primarily in clinical medicine and the social sciences. He writes on his Stanford webpage: '*Some of my most influential papers in terms of citations are those addressing issues of reproducibility, replication validity, biases in biomedical research and other fields, research synthesis methods, extensions of meta-analysis, genome-wide association studies and agnostic evaluation of associations, and validity of randomized trials and observational research.'* We all know and keep citing his famous paper *'Why Most Published Research Findings are False'.* (Ioannidis, 2005) He has since 2010 published continuously at a dazzling speed on that same subject in different fields and from different angels. He was involved in the Equator initiative and established METRICS in 2014. (https://metrics.stanford.edu/about-us) METRICS is a meta-research and innovation centre at Stanford. Ioannidis is worldwide recognized as one of the scientists with Doug Altman, Richard Smith and some others who started the debate about issues of quality and reproducibility in biomedical research. Until the COVID-19 pandemic, he travelled almost continuously around the world to deliver his passionate presentations. In the spring of 2015, I was formally invited to become a METRICS affiliate and to give a talk about Science in Transition at the METRICS inaugural conference in November of that year at Stanford Campus. As in the good old days of my AIDS research team, we had organized some visits in the Bay Area the day before the meeting. At Berkeley we met with the Vice Chancellor for Research Christopher McKee and his policy advisors and representatives of the Center for Science, Technology, Medicine & Society. A group of people involved in actions to improve the relationship between science and society. The latter were enthusiastic, the former more critical and even a bit cynical in reaction to my short pitch on how to improve science. 'You think you can change a system?' Leaving the campus, we came across a number of reserved parking spots for Noble Laureates. It was a bit weird, as we had just been talking about the skewed appreciation of different types of science. Here I have to admit, at UMC Utrecht I used to have the privilege of reserved parking close to my office which came with the membership of the board. So, I should be quiet.

At the new UCSF campus we met with Ron Vale a pioneer who was about to found ASAPbio (Accelerating Science and Publication in Biology), promoting the use of preprints and an open and transparent peer-review process. Ron did believe we can change a system. Downtown San Francisco, we discussed with Paul Volberding and his team who were involved in an interesting novel funding scheme (RAP) at UCSF to '*foster collaborative, novel, or preliminary research activity, and to further institutional research strategic goals.* Paul was in the first decades of the AIDS epidemic a well-known pioneer in the organization of clinical care, anti-viral therapy and had deeply engaged with the gay community. We only knew each other's names from these days, but still it helped to make the connection.

California Dreamin'
A bit gloomy from these encounters, we drove down Highway 1, from San Francisco, via Half Moon Bay and then through the foothills to Palo Alto. It was clear, the project of Science in Transition had a very long way to go. My colleagues Rinze Benedictus and Susanne van Weelden had carefully observed the various responses to my 'elevator pitches' in the meetings that day. They did what was to be expected from them and during the ride provided a critical analysis of the different reactions we had gotten. The higher in office the more evasive the responses appeared to be, which of course made perfect sense, given the reputational and financial interests linked up to the reward system we were all in. This did made it very clear, we were up against a major force. Fortunately, the foothills in the magical afternoon Californian light to the left and the sight of the incoming rolling clouds above the cold ocean to the right, cheered us up at least a bit. We reassured ourselves, we have to keep the ball rolling, we were doing a good thing for science and mankind. Sometimes 'on a winters day' you need these maybe naïve idealistic moments to keep you going. It's a shame though, that I could not locate the oceanfront hippie- style restaurant that I remembered, or I thought I remembered, where we liked to go during my sabbatical at DNAX in 1994.

The METRICS conference was hosted in a venue at the heart of Stanford Campus, with its bright sunny sky and the skyline of the Foothills in the background. With a lot of well-groomed outdoor sports accommodations, Stanford Campus misleadingly looks like a Spanish holiday and golf resort. At the time of my sabbatical at DNAX, then an amazing academic-style small biotech institute, the age of biotech and of internet companies just had started. I saw again how misleading and seducing the leisurely appearance of Palo Alto, like most of Silicon Valley and its people, is. In Boston, New York and Chicago, you can tell from the way the cities look and how the people in public spaces behave how tough life must be, but for some mysterious reason not so in Silicon Valley. Yet, Stanford University and the biotech and fintech companies are the engine of that most competitive region in science and innovation of the

(continued)

world. They don't show it, but people are very eager and work long hours with for most of them also long daily commutes across the bay where housing is affordable. The right place, I would say, to discuss the perversities and adverse effects of hyper-competition for social and professional credit to obtain research grants or investments from venture capital companies in order to make it to the next round. This is the world of science that Steven Shapin described and analysed in his *The Scientific Life: A Moral History of a Late Modern Vocation. (*Shapin, 2008*).*

The meeting was a warm bath, vibrant and full of positive energy. Everything and everyone was in tune, aiming in some way or another to improve the practice of science and inquiry. To be honest, at that time I did not realize how much experience, knowledge and involvement had been brought together in that meeting. There was a lot on methods, design and reporting, but fortunately the program was much broader in its approach. We also heard talks about preregistration, about animal studies and about education of representatives from patient advocacy groups, which reminded me very much of the AIDS advocacy we had seen in the 1980s. It was not only about biomedical research, as Jelte Wicherts from Tilburg University and Brian Nosek spoke about the ongoing actions with respect to reproducibility in the field of psychology. Nosek is the founder of the Centre for Open Science (COS) (https://cos.io/about/mission/) and only later I realized that Brian Nosek's talk was my first conscious encounter with the broader movement of open data, data sharing and reproducibility in the frame of Open Science. I did my 'Change the Incentive and Rewards' pitch going through the Credibility Cycle and our ongoing initiative to implement a novel evaluation system.

At the meeting Monya Baker, of the journal *Nature* had expressed her interest in our actions a UMC Utrecht and wanted to stay informed. Rinze Benedictus met with her at the meeting and kept her up-to-date. When the evaluation scheme and our CV portfolio was accepted for use at our institute, we wrote a small piece about it for *Nature* telling the story but also discussed the problems that we had seen and still anticipated for the process of implementation. We were happy with this piece since it clearly signalled that this type of action to change an important aspect of the system can be done at the institute level. The article came out in October 2016 and was picked up by Dutch newspapers, probably because it was in *Nature*, which in light of our mission was paradoxical since it said that a paper in *Nature* is not per se top class science. But anyhow, it had impact because it reached a large public. (Benedictus & Miedema, 2016). Through these international contacts we set up an exchange and collaboration with Ulrich (Ulli)Dirnagle and his team who have set up QUEST, as part of The Berlin Institute of Health (BIH). The BIH and its QUEST Center is focused on improving and transforming biomedical research, in analogy to the Science in Transition movement with emphasis on reproducibility and translational medicine but also on changing the recognition and rewards system working closely with Equator and METRICS Stanford.

6.8   Academic Rewards and Professional Incentives

At the METRICS meeting, we decided to work jointly on the problem of incentive and rewards. We focussed on the criteria applied in the career advancement system in medical schools. Steven Goodman, David Moher and I took up that task, working towards a workshop on *Academic Rewards and Professional Incentives.* David Moher and colleagues did almost all of the work. Luckily the committee working on this in my institute had delivered in the first quarter of 2016 and our evaluation scheme was included.

Change scientific incentives and rewards for broader impact & open science

REDEFINE EXCELLENCE
**Fix incentives
to fix science**

*Rinze Benedictus and
Frank Miedema*

Provide incentives and rewards
for academics to work on Open
Science and use of Open
Science

Funders (public and private)
want us to work according to
Open Science and Open Access

**BMJ Open Science, January 2018
"Who are we answering to?"**

Universiteit Utrecht

27 OCTOBER 2016 | VOL 538 | NATURE | 453

We decided to invite a select group of participants of whom most were happy to take part and could make it on Monday 23rd of January to Washington DC, just around the corner where Donald J. Trump the Friday before had been inaugurated as president of the United States.

We were most happy to welcome among the participants: Michael Lauer (NIH); Marcia McNutt (National Academy of Sciences); Jeremy Berg (Editor in Chief *Science*); Robert Harrington (Chair of Medicine, Stanford); James Wilsdon (University of Sheffield), Paul Wouters (CWTS, Leiden); René Von Schomberg, PhD (Team Leader–Science Policy, European Commission); Paula Stephan (Georgia State University); Ulrich Dirnagl (Charité –Universitätsmedizin, Berlin); Chonnettia Jones, (Director of Insight and Analysis, Wellcome Trust); Malcolm MacLeod, MD, Professor of Neurology and Translational Neuroscience, University of Edinburgh); Sally Morton (Dean of Science, Virginia Tech, Blacksburg); Deborah Zarin, (Director clinicaltrials.gov), Alastair Buchan (Dean of Med School, Oxford);

Trish Groves (BMJ, BMJ Open); Stuart Buck (Laura and John Arnold Foundation). John Ioannidis was chairing the meeting.

Guess what we talked about during the pre-workshop dinner that Sunday night. With this presidency, what had the future in store for the world, the US and US science, the EPA and the NIH? Uncertainty and great worries prevailed. The meeting was productive in the unique sense that, many different perspectives on the problems of the current practice of science were exchanged. The methodologists, the bibliometricians, Open Access advocates and the people focused on the systemic problems were in the same room.

A paper, written by Moher et al. to share the information and insights discussed at the meeting, was published early 2018. (Moher et al., 2018) The abstract of the paper was clearly a call for action:

> *Assessment of researchers is necessary for decisions of hiring, promotion, and tenure. A burgeoning number of scientific leaders believe the current system of faculty incentives and rewards is misaligned with the needs of society and disconnected from the evidence about the causes of the reproducibility crisis and suboptimal quality of the scientific publication record. To address this issue, particularly for the clinical and life sciences, we convened a 22-member expert panel workshop in Washington, DC, in January 2017. Twenty-two academic leaders, funders, and scientists participated in the meeting. As background for the meeting, we completed a selective literature review of 22 key documents critiquing the current incentive system. From each document, we extracted how the authors perceived the problems of assessing science and scientists, the unintended consequences of maintaining the status quo for assessing scientists, and details of their proposed solutions. The resulting table was used as a seed for participant discussion. This resulted in six principles for assessing scientists and associated research and policy implications. We hope the content of this paper will serve as a basis for establishing best practices and redesigning the current approaches to assessing scientists by the many players involved in that process.*

6.9   The Future of Science in Transition

In the meantime, in the spring of 2015, we decided to explore the future, if any, and we invited a few persons with visibility and authoritative in the field of science and society to join our core team of Science in Transition. Frank Huisman because of heavy duties, could not contribute anymore. These workshops were the basis for the program of the Third Symposium held in March 2016 again at the Royal Society. The problems and opportunities of the academy, the position of the PhD's, the small-scale colleges and our relationship with society and the publics were the main themes. James Wilsdon, a long-time key opinion leader in the UK and Europe was the major guest speaker who presented the findings and recommendations of 'Metric Tide'. (Wilsdon, 2016) The symposium was quite optimistic in tone, given the actions that were already ongoing in the field, but it was clear that in the next phase 'academic leadership' should step up and act. We concluded with a discussion on a typical Dutch experiment in the spirit of the Science Shops of the 1970s: The National Science Agenda. This much debated initiative from Jet Bussemaker, the Minister of Higher Education sought to invite proposals from the public about issues for scientific inquiry. As of this writing, after 16.000 proposals and allocation of the first rounds of money, the funding agency is still struggling how to deal with it and how to continue the next years. The question obviously is whether this was the right framework. Engaging the publics is critical, but that should be beyond an inventory or wish list of all kinds of questions for research (like 'why is the sky blue?'). As argued above, engagement is not to be thought of as interests at the individual level, but a social action of publics focused on problems that have social and political priority, which may change over time.

Sarah de Rijcke, photo taken by Bart van Overbeeke

From 2016, the agenda and activities of Science in Transition very much resonated with Open Science in The Netherlands and the EU were experts wrote excellent reports on the main issues related to implementation of Open Science in the member states. At several institutes in member states and around the world actions especially regarding Recognition and Rewards and use of meaningful metrics were started. The responsible administrators of universities and funders many times went for advice to experts from the field of research evaluation like Sarah de Rijcke, a senior staff member and now a full professor and Scientific Director at CWTS Leiden, who had joined our 'team' in the spring of 2015. Sarah, co-author of the Leiden Manifesto then already for many years had specialized in social studies of research evaluation and had as group leader been engaged in many large international EU projects. She had then already started a large research effort to empirically evaluate interventions in the incentives and rewards system. This included our intervention at UMC Utrecht with Rinze Benedictus as one of the PhD students. Sarah, with James Wilsdon, in 2019 established a new high-profile institute, *the Research on Research Institute*, with international support from major relevant parties. http://researchonresearch.org. Sarah is an internationally recognized expert involved in international outreach activities related to use and meaning of metrics and science management and policies. Finally, but promising for the future of science, at 'my own' UMC Utrecht, a group of four young female PhDs took the initiative to launch *Young Science in Transition* with highly relevant and visible activities: '*How young researchers can re-shape the evaluation of their work. Looking beyond bibliometrics to evaluate success.*

https://www.natureindex.com/news-blog/how-young-researchers-can-re-shape-research-evaluation-universities

References

Benedictus, R., & Miedema, F. (2016). Fewer numbers, better science. *Nature, 538*, 453–455. https://doi.org/10.1038/538453a

Bourdieu, P. (2010). *Distinction: A social critique of the judgement of taste*. Routledge.

Glasziou, P. (2014). The role of open access in reducing waste in medical research. *PLoS Medicine, 11*(5), e1001651. https://doi.org/10.1371/journal.pmed.1001651

Glasziou, P., Altman, D. G., Bossuyt, P., Boutron, I., Clarke, M., Julious, S., Michie, S., Moher, D., & Wager, E. (2014). Reducing waste from incomplete or unusable reports of biomedical research. *The Lancet, 383*(9913), 267–276. https://doi.org/10.1016/s0140-6736(13)62228-x

Hicks et al. (2015). The Leiden Manifesto on research metrics. *Nature, 250*, 431. https://doi.org/10.1038/520429a

Ioannidis, J. P. (2005). Why most published research findings are false. *PLoS Medicine, 2*(8), e124. https://doi.org/10.1371/journal.pmed.0020124

Macleod, M. R., Michie, S., Roberts, I., Dirnagl, U., Chalmers, I., Ioannidis, J. P. A., … Glasziou, P. (2014). Biomedical research: Increasing value, reducing waste. *The Lancet, 383*(9912), 101–104. https://doi.org/10.1016/s0140-6736(13)62329-6

Maister, D. H. (1997). *True professionalism: The courage to care about your people, your clients, and your career*. Free Press.

Moher, D., Naudet, F., Cristea, I. A., Miedema, F., Ioannidis, J. P. A., & Goodman, S. N. (2018). Assessing scientists for hiring, promotion, and tenure. *PLoS Biology, 16*(3), e2004089. https://doi.org/10.1371/journal.pbio.2004089

Roussel, P. A., Saad, K. N., & Erickson, T. J. (1991). *Third generation R&D: Managing the link to corporate strategy*. Harvard Business School Press.

Sarewitz, D. R. (1996). *Frontiers of illusion: Science, technology, and the politics of progress*. Temple University Press.

Sarewitz, D. R. (2016). Saving science. *The New Atlantis, 49*, 4–40.

Shapin, S. (2008). *The scientific life: A moral history of a late modern vocation*. University of Chicago Press.

Stokes, D. E. (1997). *Pasteur's quadrant: Basic science and technological innovation*. Brookings Institution Press.

Wilsdon, J. (2016). *The metric tide: The independent review of the role of metrics in research assessment & management*. SAGE.

Chapter 7
Transition to Open Science

Abstract Many initiatives addressing different types of problems of the practice of science and research have been described or cited in this book. Some were one-issue local actions, some took a broader approach at the national and some at EU level. Some stayed on, others faded after a few years. Many of the issues addressed by these movements and initiatives were part of the system of science and appeared to be systemically interdependent. This is how they converged and precipitated in the movement of Open Science, somewhere at the beginning of the second decade of this century. I discuss the major move that was made since 2015 in the EU to embrace the Open Science practice as the way science and research are being done in Europe. This elicited tensions at first foremost relate to uncertainty regarding scholarly publishing, of how and where we publish open access. But also, with respect to what immediate sharing of data and results in daily practice of researchers means, how we value and give credit for papers and published data sets. It thus poses the question of how, if at all, we must compare incomparable academic work, how we get credit and build reputations in this new open practice of science. It is indeed believed that Open Science with its practice of responsible science will be a major contribution to address the dominant problems in science that we have analysed thus far, or at least will help to mitigate them. Open Science holds a promise to take science to the next phase as outlined in the previous chapters. That is not a romantic naive longing for the science that once was. It will be a truly novel way, but realistic way of doing scientific inquiry according to the pragmatic narrative pointed out.

The Transition to Open Science as can be anticipated from the analyses above will not be trivial. The recent discussions have already shown that the transition to Open Science, even between EU member states, is a very different thing because of specific national, societal and academic contexts.

I will conclude this chapter reporting some of my first-hand experiences, in Brussels and during visits to several EU member states in the course of a Mutual Learning Exercise, but also encounters in North America, South East Asia and South Africa where we in the past years have discussed Open Science. Although we know science and scholarship have many forms and flavours and that wherever you go, there is not one scientific community. For me discussing the Transition to Open Science in the past four years was really a Learning Exercise, an amazing, mostly encouraging, but many times quite shocking, even saddening adventure.

7.1   The Big Elephant in the Board Room

In the previous chapters I have discussed the origin and history of the, in my opinion, most relevant developments in the philosophy and sociology of science. They were discussed in the wider context of changes in society in the past hundred years and how sociologists and scholars in political and social theory have reflected on them. In most cases the scholarly work was 'academic' in style and reflective about the practices of science and research and its problems. I have shown how many scholars despite the demise of the Legend, found that its legacy still had and has distorting effects on our image and the practice of science, even until this day. The fact, that there is no claim to truth based on absolute timeless foundations, for philosophers of science was hardly bearable, but as it appears, was also hard to swallow for practicing researchers in academia. But this problem goes beyond science as we saw in the Chap. 6, and what Anthony Giddens articulated in 1994: *'What seems to be a purely intellectual matter today – the fact that, shorn of formulaic truth, all claims to knowledge are corrigible (including any meta-statements made about them) – has become an existential condition in modern societies'*. Not only science but our whole everyday life *'is built on the shifting sand; it has no grounding at all'* Giddens concludes with Popper (p87) (Beck et al., 1994). Despite having demonstrated the serious distorting effects of the Legend, in our times mainly via the incentive and rewards system on the agenda and impact of scientific research, very few authors have questioned these practices at the political and organizational level in academia. Even fewer still started to propose concrete interventions to be done by responsible academic leadership to improve science and abolish these problematic practices. We have seen in the previous chapters that the reputational reward system is most likely the most critical process in academia. Almost every relevant aspect of scientific research is, directly or indirectly determined by it. The response of the establishment, that it is 'not about power and the execution of power, but all about quality and excellence' is as we have seen obvious. In defence of research, we are told that 'researchers follow their altruistic voice of vocation in search for truth, independent of personal advantage or gains'. Although most researchers, I am convinced, still aspire to that ideal, this is not a helpful defence as it blocks the attempts to make the changes to facilitate researchers to really do the research they and stakeholders from society consider as most relevant with the most relevant results and impact. The institutionalization of science as a major social system of great importance to society has however developed its own economic laws. These are inhibitory to the idealistic motivations and aims with which the individual researchers entered the field. It appeared that problem in academia that have in the past twenty years been exposed in analyses of various movements -open access publishing, data sharing, public engagement and outreach, poor reproducibility and waste- cannot be properly addressed and solved without taking on this problem of the system. As said before, it is the 'Big Elephant in the (Board) Room'. Only by taking the systems approach and its corresponding interventions, we are able to gradually, but fundamentally change the practice of science by which the different actors in the field are incentivized and empowered to 'do the right science right'.

The movement of Open Science as it has come of age in 2020, aims to truly integrate concrete actions that take on virtually all of the problems of science that have been revealed by previous analyses and movements. In 2016, the EU explicitly adopted Open Science, including the change of the indicators used in the practice of Incentives and Rewards. Like Science in Transition, '*Equator/Rewards*' that came from the Lancet '*Reduce Waste, Increase Value*' initiative by an international consortium in collaboration with *Lancet,* clearly since the start in 2014 have engaged with the ideas op Open Science, as has the Meta-Research Innovation Center at Stanford (METRICS).

In this chapter, I will briefly discuss the major movements that can with hindsight be regarded, in one way or another, as preludes to Open Science, as each of them has focussed on different issues from different scientific and societal perspectives. I regard the Responsible Research and Innovation program critically important as groundwork to make the full-fledged adaptation of Open Science by the EU in 2016 possible. I realize that this may not be a generally shared perspective and recollection of the developments at the EU DG Research and Innovation. In my opinion the EU Open Science program worked on the technical issues enabling Open Access and Open Data, but these were means to an end. The program aimed for an optimal and open relationship between science and academia and the various stakeholders in society for which Open Access and Fair Open Data. It also integrated in the Open Science Program a program on the required change in the reward system. The EU Open Science Program did not look away from that elephant in the room. I will discuss the Open Science movement as it has been developed since 2016 in the EU and elsewhere in the world. I will refer to the present-day use of Open Science practices in the heat of the COVID-19 pandemic, but also duly pay attention to concerns regarding some practices of Open Science. Finally, the promise and future of Open Science will be discussed in light of recent geopolitical developments in which the USA, China, but also the EU are re-thinking their science and technology strategies.

7.2 Responsible Research and Innovation

Brussels in the Meantime

Jan Staman, the Director of the Rathenau Institute, invited me to give a short presentation about Science in Transition at a meeting in Rome in September 2014. The meeting was about an EU project with the title *Responsible Research and Innovation*. I had recently been in a couple of public debates with Staman who was very supportive about our initiative. For Staman, a veterinarian trained in Utrecht, the relationship between science and society was not only real, but urgent. This was fuelled by the recent outbreaks of SARS, MERSH and Q fever, all caused by zoonotic pathogens leading to serious public health problems when they jump from animals to humans. In my quite impolite one-line reply, I was very blunt to say that I had no idea what RRI was about, but that I was interested to spread our message on an EU podium. This, I now realize, must have hurt Jan Staman. After twelve years he was just about to leave his job as director of Rathenau handing over to Melanie Peters from Utrecht University, who we knew well from her interest in Science in Transition. In a fare-well interview in a national newspaper, he complained loudly that Rathenau had a hard job engaging the elite institutes and scientists to do research on the grand societal challenges and that this was a battle that had been going on for more than forty years or so. He was mild about the Academic Medical Centres and the Technical Universities who were, he said, closer to societal problems. He apparently had been too loud and got backfire from Carel Stolker, the Rector of Leiden University and Hans Clevers, the President of the Royal Academy. They argued that this was a caricature of science, since many researchers were very engaged in societally relevant research. Indeed, there are those who are, but is it respectable, is it facilitated and do we reward them enough? For most the issue still is the response: 'The ERC, yes, but must we really engage with these large, messy less-focussed problem-driven consortia in Horizon 2020?'

In Rome, I admit, I was embarrassed that I had not noticed and researched RRI before. There were not the fifty people I had expected, but more than thousand people in the meeting with impressive talks and lively sessions about major EU programmes and investments amounting to literally hundreds of millions of euro's and 400 million to come until 2020 in actions on Public Engagement, Diversity and Open Science. The closing talk was to be given by Bryan Wynne, whose work I introduced in the previous chapter. How could it be that I did not know this highly relevant program in which major key opinion leaders in European STS and of the movements of citizen science and even already Open Science were involved? It did not seem to be uniquely my problem. The overall penetration of the RRI movement in academia was low. There was, however, little to be found in terms of analyses why this EU

(continued)

programme, after many years and major investments, still was not mainstream policy in academia. There was no systemic organizational bottle neck identified and hence there was no action plan for academic leadership to make the required change. This program would, as it was, not become mainstream and would not bother the 'high church' too much. That was exactly my pitch at the end of the second day, just after I met briefly during the coffee break with Arie Rip, one of the key players of STS since the 1980s and one of the founders of The Rathenau Institute.

The roots and development of *Responsible Research and Innovation* (RRI) in the EU, from 2000 on, have been adequately described (Owen et al., 2012) (Stilgoe et al., 2013; René von Schomberg & Hankins, 2019; ESF, 2013). RRI stems from a series of different initiatives to increase integrity, ethical, legal and social responsibility and to intensify multidisciplinary research to integrate social science with technical sciences and innovation. Programs preceding RRI where of the type discussed in Chap. 5, on public participation and deliberation with a theoretical perspective but also based on cases studies of problematic issues like GM crops, ICT and genetic engineering and on 'real time' technology assessment. I referred already in Chap. 5 to the work of Wilsdon, Owen, Wynne, Irwin, Felt, Stilgoe, Rip, von Schumberg and Sarewitz and their colleagues in the first decade of the century. Here and there in these studies, open innovation an openness to the public is mentioned as a tool to enhance impact. These authors are strongly in favour but share concerns about responsible research and development – the design and introduction and use of innovations- with respect to collaborations with private and commercial partners. They also are cautious of the problematic and most often unanticipated social and economic effects upon implementation of technology. Most of them do argue for up-stream participation by stakeholders in the knowledge production process, which was in some fields, most prominently in medical research, already being used but mostly not in an institutionalized way.

As Felt et al. have shown graphically, (ESF, 2013)(p11), these flavours of RRI were already visible in the programmes of the EU between 2000 and 2013. In RRI, through the practice of Knowledge Transfer and Public Engagement, societally responsible research and innovation requires a broad and deep understanding of its ethical, legal and social implications (ELSI) or aspects (ELSA). For a thoughtful series of papers on management of RRI, with emphasis on these responsibilities of the different parties involved, I refer to a collection of papers by experts in *Responsible Innovation* (Owen et al., 2013). They analyse in detail the public debates about innovation in nanotechnology, geoengineering, information technology (AI) and finance.

In 2011 the EU took the initiative to unite these movement and ideas under the banner of RRI as part of Horizon 2020, the framework program for 2014–2020. Owen et al., describe this process and point to an influential paper by René von Schomberg (Owen et al., 2012; Rene Von Schomberg, 2011). René von Schomberg at the time of this writing still is a thought-leading and senior policy advisor at the EU DG Research and Innovation. Rene von Schomberg's paper was circulated and was crucial in this development because it was according to Owen *'outlining his emerging philosophical thinking, …*(that*) included a thoughtful discussion concerning the normative targeting of research and innovation towards the 'right impacts…'*. This was science aiming at economic, but also health and social problems, based on external social and political values and goals which were broadly expressed in the Treaty of the European Union. Von Schomberg in the paper discusses that in our times the Aristotelian concept of 'the good life' as the purpose of science may be problematic, but that missions and challenges defined in the debates and the deliberations that are found in the EU treaty can give normative guidance. He provides philosophical depth, how research that has been brought in the context of a public controversy is being analysed and deconstructed. In that interaction, the debate is often, not about the concrete claims of research, but about which type of research is best suited to be taken in to account in a specific social and technological controversy. In addition, the problem for science is that while an epistemic debate (about scientific knowledge) is going on and not yet closed, it has induced, or fired up public debate. *"Which group of scientists can we believe, and should we endorse? Plausible, epistemic approaches on the acquisition of knowledge in science are associated with problem-definitions, which in turn frame (although, often, only implicitly) policy approaches."* He argues for a strong science-policy interface which allows for *'deliberation based on normative filters such as proportionality and precaution'* with respect to societal intervention or actions which are EU principles (Rene Von Schomberg, 2011).

We have seen in the previous chapters in the cases described by Wynne and Irwin how this asks for reflexivity from the researchers and obviously adds complexity to the process of policy making. Von Schomberg writes explicitly about the issue of problem choice and its coupling to research policy and investment and in a later paper: *'Under the European Framework programme for Research and Innovation Horizon 2020, a number of 'Grand Societal Challenges' have been defined, which followed the call in the Lund Declaration for a Europe that 'must focus on the Grand Societal Challenges of our time' (Lund Declaration 2009 during the Swedish EU presidency). Sustainable solutions are sought in areas such as "global warming, tightening supplies of energy, water and food, ageing societies, public health, pandemics and security. Arguably, the Grand Societal Challenges of our time reflect a number of normative anchor points of the Treaty in relation to the 'promotion of scientific and technological advance' and which thus can be seen as legitimate. However, the promotion of scientific and technological advance has until now served as a goal in itself. The promotion of scientific and technological advance has not been coupled to other, all interrelated, normative anchor points such as 'ensuring a high level of protection' that, 'sustainable development', 'competitive social market economy' that drive all other EU policies. It does not require much political initiative to couple the promotion of scientific and technological advance with all other major normative anchor points in the EU treaty to give a broader base for the justification of research and innovation beyond assumed economic benefits and increase of competitiveness.' (René von Schomberg, 2019).*

RRI included science for society with early participation by the public, acknowledging all these complexities. It is about science with society in which the relationship with society was integrate and institutionalized such that it could be anticipated, reflected upon and be opened up to the diverse stakeholders and publics. Owen et al. emphasize that this *'confers new responsibilities: and not only on scientists but universities, innovators, business, policy makers and research funders.'* This regards to program choice and responsiveness to their delivery. Owen et al. state *that 'The framing of responsibility itself is perhaps one of the greatest intellectual challenges for those wrestling with responsible innovation'.* How can you deal with that in issues where high risk and high uncertainty is involved? This asks for reflection on the goals of research and innovation and a reflexive mode of research that is responsive to all kinds of social impacts that it will bring or has brought about. Obviously, this demands more inclusive codes of conduct, research ethics and scientific integrity. This is quite different from the classical idea of the Legend that scientists produce neutral knowledge which can in the next stage be translated, applied and used either to the good or bad causes for which the scientists feel they cannot not be held responsible. The *Rome Declaration on Responsible Research and Innovation* stemming from the EU program is reproduced that provides a clear overview of the program (Supplement 5). Almost all of the authors writing about RRI and mentioned above had European affiliations, so it seemed logical that in the EU the next step was going to be Open Science. That is with hindsight, because when the EU launched Open Science in 2016 this was for many still a surprise, but a pleasant surprise.

7.3   The Early Voices of Open Science

In the preliminary phase, before the different movements that aimed to improve science and research were organically brought under the banner of Open Science, we have seen several important movements that with hindsight each have had major effects. In the late 1990s, the field gradually became aware of what librarians called the 'serials crisis'. Subscription prices of scholarly publications were increasing much vaster than inflation. This was going on to the effect that even in the developed countries and at well-endowed institutes, librarians to stay within their allocated budgets, had to selectively stop subscriptions.

When I started as research director of Sanquin Research in January 1998, this was one of the problems that was waiting for me. The institute was an independent non-profit foundation with a small research division and limited internal funding. Given the yearly financial pressures of the publishers, it felt logical to modernize the library. The library, as elsewhere, thus changed to digital subscriptions, with less physical librarian support. Fortunately for me this coincided with retirement of a librarian, that however did not reduce the reading costs of the journals. On the contrary, they were growing every year. So, we had to stop subscriptions based on the interests of the researchers. Later, as the dean at UMC Utrecht, I saw how this this dossier had developed even further in the same manner. The emphasis on the 'better' journals, dropping subscriptions of the 'lesser' journal, started a vicious cycle of increasing prizes of the 'better' journals who were in high demand, since the researchers appeared to be addicted to them. The higher the JIF, the more dramatic the addiction, the higher the subscription prizes. The publishers know how to play the game and offered package deals of subscriptions in order to sell also their serials who are in lesser demand. This happened not only for the publications of the 'Big Five' (Suber, 2012), but also for journals published by the so-called learned societies where these profits were used to fund their scientific activities.

Some visionary scientists already in 1991 sensed this problem and, like the publishers, taking advantage of the novel digital developments, started arXiv.org a repository for STE and economics, where researchers could publish their work for all free to read, fully open access, before it is submitted to a journal. In 2006 the Public Library of Science (PLOS) series started that published papers that are reviewed, are free to read, but ask the authors to pay Article Processing Costs (APC). In 2013 bioXiv.org, a repository for biological sciences and in 2019 medRxiv.org for biomedical sciences was launched. Repositories can be institutional or disciplinary in nature. In times of COVID-19 all research was immediately made available through repository publishing, an obvious thing to do.

The best-known movement within Open Science, no doubt, is the Open Access movement. Open Access formally started with the Budapest Open Access Initiative in February 2002, the Bethesda Statement on Open Access Publishing in June 2003, and the Berlin Declaration on Open Access to Knowledge in the Sciences and Humanities in October 2003. For a detailed and thoughtful analysis, I refer to Peter Suber who wrote a concise book as an introduction (Suber, 2012) followed up in 2015 by his vast collection of blogs in *Knowledge Unbound.*(Suber, 2016) There also is the excellent Wikipedia site and Peter Suber's own personal webpage.

We learn from that reading that interesting, stand-alone initiatives and actions have been taken place already a long time ago. These initial actions have slowly resulted in more recent actions to make research papers and data openly available. They were still sometimes local, but now are mostly national and institutional in nature. These actions were inspired and made possible by the world wide web and the possibility to read journals 'electronically'. Since the year 2000, I have not held in my hands one of the journals that I as an active researchers physically, in hard copy, used to browse in the library every Monday afternoon since 1979. Now Scientific papers can be assessed everywhere. For our kids this is the new normal, but in the 70s and 80s one still had to go to the library to browse the contents of the journals and take a Xerox photocopy of articles of interest. The well-known space limits in printed journals required deletion of experimental data which editors used to impose on authors, but that could now be more easily allowed as supplementary data. This access was, for almost all journals only available to those who could afford the subscription fees, that were steeply rising, despite its scale up in reaching libraries in the word-wide electronic markets. Already in the late 1990s some journals made themselves open, readable for free on the web, and somewhat later the first Open Access journals, like the Public Library of Science (PLOS) series, started that are free to read, but do ask the authors to pay Article Processing Costs (APC). Because of these partial technical and financial solutions and the JIF game explained in Chap. 3, it took a long time for Open Access to reach the level of penetration that it has obtained in Europe and around the world in 2020.

Another important initiative that many have heard of and that logically started from the digitalization of science and society is related to Open Data and Open Code. Among the many advocates of this movement which sometimes was designated as Science 2.0, in analogy to the participatory Web 2.0, I like to mention Michael Nielsen, a remarkable quantum physicist, science writer, and computer-programming researcher whose book *'Reinventing Discovery, The new era of networked science'* had much impact (Nielsen, 2012). Nielsen has been a scientist/ activist for Open Access and Open Science in the early years before he published the book and left academia to pursue his own projects. Nielsen has been a scientist/ activist for Open Access and Open Science in the early years before he published the book and soon after, he left academia to pursue his own projects. Nielsen shows how scientists together, but also on collaboration with non-scientists, have used the internet to solve problems, to collect and exchange data in, an in principle, world-wide digital space. He discusses the Open Access actions and in addition gives examples of how a new way of doing science and discovery work, as a networked science, has been applied already to many different problems in different fields of science and society. He mentions theoretical work on mathematical problems and work by the Centres for Disease Control in the US on influenza epidemics, which for the reader in 2020 is already quite normal.

At the time of writing in the COVID-19 pandemic, we experience the power of this networked research on a daily basis by which via different platforms data is being shared immediately in order to inform policy making around the world. He makes a strong case for networking and data sharing and concludes for that Open Science involves a cultural change for science and scientists that is seriously inhibited as it is in the old system not incentivised and rewarded (p6–8; p187–197). That these networks can be truly open is illustrated by the story of Hanny van Arkel, a 27-year old Dutch schoolteacher with an interest in cosmology who got engaged in an effort to characterize galaxies which involved 200.000 volunteers. One day in

2007 she spotted a blue bob on a photograph of Galaxy Zoo which later the scientists concluded must be a quasar mirror (p129). For Nielsen digitization is a tool which makes science open and more democratic. He is passionate about the contribution that science can have to society and hopes that this Networked Open Science way of discovery can help us to close the 'ingenuity gap', he mentions the dangers of HIV/AIDS, proliferation of nuclear arms, bioterrorism, shortages of water and oil and the effects of climate change (p171). Obviously, in the summer of 2020, the dangers we think of are COVID-19 and the pandemics to come, the immense refugee problems caused by local wars and its disasters and the social and economic problems caused by increasing global economic and social inequality.

Recently Bernard Rentier published a handy and informative overview of Open Science (Rentier, 2019). A very informative collection of papers about Open Science also is *'Opening Science, The Evolving Guide on How the 'Internet is Changing Research, Collaboration and Scholarly Publishing'* (Bartling, 2014). Both are published Open Access. In the chapter written by Fecher and Friesike in the latter book, Five Schools of Thought are presented which each combine specific aims and the tools to achieve these aims (see Table and Figure) (Fecher & Friesike, 2014). In the 'fifth school' the need for a change in the practice of research evaluation is emphasized, taking into account the typical academic activities of Open Science. Friesike has recently published commentaries in Nature, Science and an LSE blog on these issues and in his subsequent studies provided ample evidence that the individual system of academic reputation and reward is the reason why researchers in many different fields do not to practice open access and data sharing, despite its benefit to the science (Table 7.1).

Table 7.1 Five Open Science schools of thought

School of thought	Central assumption	Involved groups	Central Aim	Tools & Methods
Democratic	The access to knowledge is unequally distributed.	Scientists, politicians, citizens	Making knowledge freely available for everyone.	Open Access, intellectual property rights, Open data, Open code
Pragmatic	Knowledge-creation could be more efficient if scientists worked together.	Scientists	Opening up the process of knowledge creation.	Wisdom of the crowds, network effects, Open Data, Open Code
Infrastructure	Efficient research depends on the available tools and applications.	Scientists & platform providers	Creating openly available platforms, tools and services for scientists.	Collaboration platforms and tools
Public	Science needs to be made accessible to the public.	Scientists & citizens	Making science accessible for citizens.	Citizen Science, Science PR, Science Blogging
Measurement	Scientific contributions today need alternative impact measurements.	Scientists & politicians	Developing an alternative metric system for scientific impact.	Altmetrics, peer review, citation, impact factors

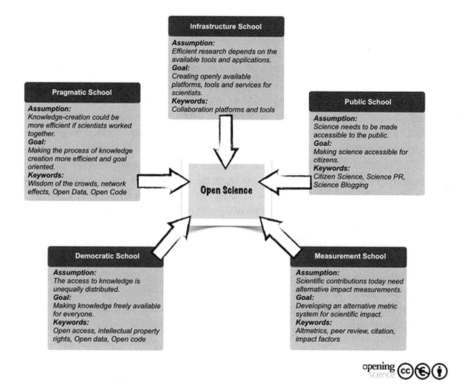

7.4   Politics, Policy and Open Science

'Biting the bullet:…the practice of having a patient clench a <u>bullet</u> in his or her teeth as a way to cope with the extreme pain of a <u>surgical procedure</u> without anesthetic' (wikipedia)

The EU and the Dutch government do not and did not always agree, but they completely agreed on the promise of Open Science and on actions to make the transition. As discussed in Chap. 3, the Dutch ministers of Higher Education and Science responded very positively to the Science in Transition initiative. This was reflected in the Science Vision of November 2014 by policies on Open Access and Open Data and renewed emphasis of the interaction with citizens and the public. This ran in 2014 parallel to the *Science 2.0 Science in Transition* initiative of the EU DG Research and Innovation. The latter started with a background paper for a survey to get a feel for the ideas and problems of science in the field of the various stakeholders. As mentioned in Chap. 3, this may be considered the prelude for Open Access and Open Science in the EU (Burgelman et al., 2019).

In the fall of 2015, the ministry of Education, Science and Culture (OCW) began work on the agenda for the first half of 2016 when the Netherlands was to hold the Presidency of the Council of the European Union. For science and innovation, the emphasis was on Open Access and a better relationship of science with politics and public to enhance innovation, and economic growth. These items were put in the larger context of Open Science that came out of the EU Science 2.0 project. The larger Open Science framework has a lot of overlap with that of Science in Transition and joining forces was logical. For Science in Transition, Open Access (OA) was believed to be relevant but was regarded as mainly a technical problem of the organization of academic scholarly publishing. We, perhaps a bit naively reasoned that it would be automatically (*en passant*) solved when we adopted DORA to get rid of the 'impactfactormania'. The most important thing that would promote the widespread implementation of OA thus for sure was the simultaneous change in incentives and rewards. The reasoning is that open access journals that are totally open and have no subscription costs have a lower JIF compared to the classical 'top' journals that have steadily and consciously build their reputation. So, as long JIF's still are overvalued and dominantly used, scientists don't like to publish OA. For sure, making authors or their institutions paying extra to make a paper OA in *Nature* or *Cell* is not the way to solve the problem, if alone because that this would be double dipping, paying the publishers twice. The latter is broadly recognized, the idea that we needed to change research evaluation criteria however was not a general awareness, or as we have seen in the previous chapters, simply thought of as a political 'no go area'.

Getting his attention

One of our staff members, who was into national politics, introduced me to Sander Dekker, the State Secretary for Science who was leading the Science and Innovation theme in the program for the Dutch EU Presidency the first half of 2016. In November 2015, I had the opportunity to talk for an hour with Dekker at the end of one of his many busy days. He is a sociologist by training and curious and eager, so when I opened my laptop and walked him through Bourdieu's credit cycle in which JIF is 'the real thing' and Open Access thus is nice the have at max. It was immediately clear to him that the problem of incentive and rewards should be part of the '*Amsterdam Call for Action on Open Science*'. In January 2016 there was a meeting in Brussels organized by the Dutch ministry to prepare for The Presidency Conference in Amsterdam where the agenda for Open Science for the EU was to be drafted. The meeting showed a for me unanticipated enthusiasm and drive for actions to make the transition to Open Science among the participants of the EU offices in Brussels, LERU but also from several members states. There were two breakout sessions on incentives and rewards, but also about research infrastructures needed to facilitate data sharing. In the following months I was invited to make the case for changing incentives and rewards, based on our UMC Utrecht pilot, at the EU Presidency Conference held on April 4 and 5 in Amsterdam.

At the closure ceremony of that meeting a preliminary draft of the '*Amsterdam Call for Action on Open Science*' was presented to Sander Dekker and Robert-Jan Smits, Director-General of DG Research and Innovation (RTD) of the European Commission. The plan was comprised of five action lines that focus on open access to publications and optimal re-use of research data, but also on necessary changes within the science system in order to attain a new and sustainable situation with respect to an open science system. I still tend to believe that this call, although of course very much a symbolic act of the Netherlands Presidency and of the EU, has been a major step in the transition to Open Science in Europe and beyond. As the EU is a major factor in global science, one may expect and hope that it may eventually turn out to be an important action for the global transition to Open Science. (Supplement 4) This Action Plan, which was based on a Draft Agenda published two months before, makes it very clear that with Carlos Moedas in his role of commissioner and main political figurehead, the EU was going for Open Science with everything that had to come with it. In this movement, the EU was going to proverbially '*bite the bullet*' at least two times. First by proposing to reform the incentive and reward system (Action 1, shown above), and second by taking actions to change the system of scholarly publishing (Actions 4, 7–10). The other actions, surely where brave and would also require major efforts but were not thought to meet with the resistance from the academic institutions that Action 1 might experience. This was the ambitious EU Open Science agenda for the years to come and in fact it had already had a flying start in Brussels. In the course of 2015, Carlos Moedas, the Commissioner for Research, Innovation and Science had already given a couple of visionary talks in which he outlined the Open Science program of the EU. It was, at least as it looked to me, to me based on the RRI programmes, now put in the perspective of Open Science. The full narrative of these preliminary messages was published in the book '*Open Innovation, Open Science, Open the World*' that was written by a collective of authors from DG R&I at the end of 2015 and formally published by the EU in May 2016 (EU, 2016). The classical narrative of entrepreneurial science and innovation in open collaboration with major partners around the world, in this agenda was put in the frame of Open Science.

For the Open Science movement, in Europe but also in the world, this in my opinion was a truly historic moment. This program did put the by now well-known issues of Open Access and Open Data in a much wider conceptual and science-policy frame. It explicitly advocated a different way to do science and research in a truly co-operative open and responsible relationship with society. You could see it as a movement to fully embraced the RRI program and transform it to the top level of EU science policy. Open Science was to be the founding principle of EU research and Innovation. It was the declaration of **'the way how we do science in Europe'** with emphasis on fruitful interactions in the different societal contexts. Experts recognized the ideas of 'well-ordered science' and deliberative processes in modern democracy.

7.5 EU Stakeholder Consultation on Open Science Policy

The transition to Open Science and research, as it has also been termed, was a change to the mainstream practice and would require complex systemic changes which involved cultural-behavioural interventions as well as infrastructural solutions. It was foreseen to have a number of Expert Groups giving advice to the Commission on issues for which advice was thought to be badly needed. The eight policy ambitions that needed to be addressed in line with these five broad action lines.

1. **FAIR open data**
2. **European Open Science Cloud**
3. **Altmetrics**
4. **New business models for scholarly communication**
5. **Rewards**
6. **Research integrity**
7. **Open science skills**
8. **Citizen Science**

In 2016 already two of these Expert Groups had been started, one on Altmetrics and one on Rewards. Fortunately, they appeared to have already broadened their tasks to problems of rewards and research evaluation when they reported in the spring of 2017. In *'Next-generation metrics: Responsible metrics and evaluation for open science'*. James Wilsdon and colleagues, among whom Paul Wouters, discussed not only the problem of the abuse of metrics but also the broader criticisms of recent scholars and movements and recommended the development of responsible metrics to incentivise and reward the practices of Open Science to come to a more inclusive evaluation of results of academic work.(EU, 2017b).

Through 2019, Paul Wouters chaired a second Expert Group to further delve into the problem of research indicators for Open Science, providing a broad approach with room for freedom in the choice of indicators and room to develop more appropriate indicators dependent on the widely differents contexts of the research. They appropriately did take into account that indicators, to the disappointment of some higher management, often are incomparable because very much dependent on the research contexts of the respective fields and sub-fields. Interestingly, clearly showing the theoretical and practical experience of the group, they called for cautiousness when implementing new indicators, warning for unintended harm they might cause to the practice of science (EU, 2019).

The Expert Group on Rewards started in July 2016 with the following task:

1. Promote a discussion with stakeholders on the current reputation system in the context of the standing ERAC groups and the Open Science Policy Platform (OSPP) which will work on the concretisation of a European Open Science Agenda;
2. Within the OS environment, reflect about and propose alternative methods to recognise contributions to OS, including 'rewards and incentives' taking into account diversity in experience and career paths, while guaranteeing fair and equal career development of individual scientists;
3. Propose new ways/standards of evaluating research proposals and research outcomes taking into consideration all OS activities of researchers, possibly recommending to pilot them under certain calls of Horizon 2020;

4. Identify existing good practices on how OS issues are already taken up by researchers, research performing institutions and research funding institutions in Europe.

This Expert Group reported in July 2017 its advice on *Indicator frameworks for fostering open knowledge practices in science and scholarship* (EU, 2017a). It is written from the perspective of University Human Resource Management and prominently features a set of indicators that can guide career assessment evaluation. Interestingly, the working group took this to a broader perspective and emphasizes more inclusive and behavioural aspects, as team science and leadership. A thorough analysis of current evaluation practices of researchers was presented, including an adequate discussion of the recent critiques, including JIF and DORA, and of some early pilots on better measures. The 'Open Science Career Assessment Matrix (OS-CAM)' presents a range of evaluation criteria for assessing Open Science activities', a practical overview that should be taken into account when evaluating scientists for using and applying Open Science practices in their research. This OS-CAM has since then been well received and propagated since.

> **Open Science on Tour in the EU**
> In a videocall with staff of DG Research and Innovation, beginning of February 2017, I was approached to chair an MLE on Open Science, especially focused on Incentives and Rewards. In the call were present the three experts who were going to take part, but also René von Schomberg, and the persons from DG R&I who were going to organize the MLE. It was explained what an MLE was all about and what was expected from us in the coming ten months or so. I had never been involved in committees or working groups of the EU, but I thought it was going to be a fascinating exercise and we all agreed to go for it.
>
> An MLE, a Mutual Learning Exercise, appeared to be a project to support member states at **"improving the design, implementation and evaluation of R&I policies"**. It appeared that nine members states had shown interest and a relevant and stiff program had been laid out already by the staff at DG R&I. The team consisted of: Katja Mayer, Rapporteur and Expert; Sabina Leonelli, Expert; Kim Holmberg, Expert; and Ana Correia, DG RTD-Unit A4. (Analysis and monitoring of national research and innovation policies); Rene Von Schomberg, DG RTD- Unit A6. (Data, Open Access and Foresight); Irmela Brach, DG RTD- Unit B2. (Open Science and ERA Policy) and Nikos Maroullis, from Technopolis for support.
>
> Being a novice in the field of Open Science and Altmetrics, I at that time only vaguely knew René von Schomberg, whom I had met just the month before at the METRICS meeting in Washington. After the first meetings I already knew somebody had done a great job at selecting the experts. My teammates were excellent and very experienced experts who were used to deliver high quality work on time. They were three scholars in Science and

(continued)

Technology Studies with quite different scientific backgrounds: Sabina Leonelli (Exeter), winner of the 2018 Lakatos Award in the Philosophy of Science for her book *Data-Centric Biology: A Philosophical Study* (2016); Katja Mayer (Vienna) an experienced social science researcher affiliated with several institutes on science, technology and policy and Kim Holmberg (Turku) an expert on scientometrics, social media and altmetrics. Sabina and Katja were energetic, outspoken and totally focussed, Kim in true Scandic Style, took his time to reflect before speaking adequately and in a low voice. With this team and the participants from several member states we met several times in Brussels and went on tour to Helsinki, Dubrovnik and Zurich to learn what Open Science would mean for the different science systems in the member states.

Topics/Sessions of the MLE on Open Science

The main topics that would be discussed are described in this section (the topics are labelled A, B, C and D in the remainder of the document). Please note that these topics may be organised differently based on the feedback from the participants during the *kick-off meeting*, and of the experts whose services are requested in this document.

Topic A: Different Types of Altmetrics

Identify and discuss different types of altmetrics that are being used or developed by universities or research funding bodies. The aim is to explore new ways/standards of evaluating research proposals and research outcome taking into consideration all Open Science activities of researchers. Evaluation criteria should take due account of the engagement of researchers in Open Science.

Topic B: How to Use Altmetrics in the Context of Open Science

Identify and discuss practical examples/best practices of how altmetrics is being used for evaluating research and rewarding researchers for engagement with Open Science The aim is to review/assess the current reputation system and adapt researcher career reward systems for engagement with Open Science practices.

Topic C: Incentives and Rewards to Engage with Open Science Activities

Identify and discuss 'good' practices for incentivising and rewarding researchers to engage with open science activities. The aim is to credit activities which are important for Open Science, such as open review and evaluation, as well as citation, curation and management of research data.

Topic D: Guidelines for Open Science

Review current state of play and share experiences in developing and implementing national policies and related actions for incentivising researchers and research institutions to engage with Open Science. The aim is to contribute to the ongoing discussion on whether/which/how common Open Science principles and requirements could be set up to affect the roles, responsibilities and entitlements of researchers, their employers and funders.

In contrast to the Expert Groups were the experts wrote one paper, in 'our' MLE the experts did write papers, concurrent with and following our discussions with representatives from the member states at the meetings held in Brussels, Helsinki, Dubrovnik and Zurich. From these conversations, speaking notes were taken that are still accessible as background information. These documents are very rich in that they demonstrate opportunities, inhibitions and caution about Open Science. In general, and in principle the attitude of the MLE participants was very positive but they very clearly pointed out the resistance and problems they anticipated. This informed us what type of action and support from the EU they would be needing in their country. They at least needed a clear 'unisono' voice from the different DG's of the Commission and the Commission that this was going to happen because it was a necessary intervention if science was to really contribute to the grand social and economic challenges.

The MLE final report, adopted by the EU Open Science Policy Platform (EUOSPP) and became part of its integrated advice to the Commission, in the spring of 2018 (EU, 2018). In the Supplements section, I reproduce the MLE Summary Article of January 2018 of which Katja Mayer was the main author.

I refer to the MLE Open Science website where all information about the MLE in a very handy format is findable, accessible and downloadable. These products written by Sabina Leonelli, Kim Holmberg and Katja Mayer reflect the way the MLE has been working, covering nearly all aspects of Open Science in explicit discussions regarding implementation, monitoring and evaluation. https://rio.jrc.ec.europa.eu/policy-support-facility/mle-open-science-altmetrics-and-rewards

In these ten months, we were discussing in depth the cultural changes required for transition to Open Science, defined much broader than Open Access. By doing so, we discussed the way the science systems in the respective member states were organized and how they would be able to adopt Open Science. The differences in academic culture were amazing and highly relevant to the topic. We were introduced, to the different path-dependent, histories and evolutions of science, in which the legacy of national political history, religion, the effects of WWII, and the Balkan wars in the 1990s, could be clearly distinguished. Differences in opinion about some of the practices of Open Science could only be fully understood after we were explained the deeper socio-economic politics of the country at informal evening diners of the country visits. We were made aware that in some countries the ministry appoints professors at the national level and research evaluation and its criteria are determined by the ministry. Some countries, totally understandable to avoid potential nepotism, just had decided to use 'objective' indicators as JIF and h-index, other countries just decided to leave such use of metrics behind and go for narratives, interviews and peer review. The most prominent example maybe the fact that after WWII autonomy of scientists has been safeguarded in Article 5 of the Basic Law of the Federal Republic of Germany, which by some implies that scientists have full autonomy to **not** engage in Open Science. These cultural differences that exist even within the EU, make you wonder how Open Science will be received and what is needed to have it adopted in China and Russia, India, African and Latin and South American countries.

Open Science – enabling systemic change through mutual learning

Small fixes are not enough to reach Open Science's full potential. Systemic and comprehensive change in science governance and evaluation is needed across the EU and beyond, report experts in a recent Policy Support Facility mutual learning exercise.

As a truly global movement, Open Science strives to improve accessibility to and reusability of research practices and outcomes. But the benefits of Open Science touch almost every aspect of society, including the economy, social innovation, and wider sustainable development goals.

"Open Science is more than Open Access and Open Data; it is a way of looking at the world, with the intent of building a better society."

Bart Dumolyn, Policy Advisor on Open Science and Responsible Research and Innovation for the Flemish Government

In its broadest definition, Open Science covers Open Access to publications, Open Research Data and Methods, Open Source Software, Open Educational Resources, Open Evaluation, and Citizen Science. But openness also means making the scientific process more inclusive and accessible to all relevant actors, within and beyond the scientific community.

With its many initiatives and programmes, Europe has long championed Open Science practices as a powerful means and excellent opportunity to renegotiate the social roles and responsibilities of publicly-funded research – and to rethink the science system as a whole.

The Horizon 2020 Policy Support Facility (PSF) gives Member States and Associated Countries the opportunity to request and take part in mutual learning exercises (MLE) addressing specific research and innovation policy challenges. The transition to Open Science represents such a policy challenge which is best tackled in close cooperation with all stakeholders and on an international scale.

Given that there is no common baseline for how to implement Open Science nationally, the MLE embraced a hands-on, 'learning by doing' approach supported by external expertise. Concrete examples, models, best practices and knowledge exchanges fostered broader understanding of the implications and benefits of Open Science strategies.

Problems and concerns were discussed in an 'open' and constructive fashion. The final PSF report, entitled 'Mutual Learning Exercise on Open Science: Altmetrics and Rewards', builds on this rich exchange of experiences, both positive and negative, and provides an overview of various approaches to Open Science implementation across Europe, which include different stakeholders and research communities.

MLE participants agreed that small fixes are not enough: implementing Open Science requires systemic and comprehensive change in science governance and evaluation. Crucial for a successful transition to Open Science will be strategic shifts in the incentives and reward systems.

"There can be no mission-oriented approach to research and innovation without Open Science."

Michalis Tzatzanis, Austrian Research Promotion Agency (FFG).

Key lessons on the transition to Open Science

The scope of this first MLE on Open Science was narrowed down to address three topics, all of which are key elements of the European Open Science Agenda:

1. The potential of altmetrics – alternative (i.e. non-traditional) metrics that go beyond citations of articles – to foster Open Science

2. Incentives and rewards for researchers to engage in Open Science activities

3. Guidelines for developing and implementing national policies for Open Science

Many MLE participants voiced concerns that altmetrics may encourage a business-as-usual scenario, with users focusing only on what is measurable and ending up with proxies far too simplistic for decision-making. Generally it was agreed that altmetrics have the potential to foster a major shift in the way research activities are evaluated and rewarded, providing they are open and reproducible in their method and data, as well as clearly indicate what qualities they measure.

So, what research qualities and societal benefits matter the most, how can they be tracked and measured, and for what reasons? Altmetrics can only help to break away from traditional indicators and publishing avenues, and establish themselves as responsible metrics if they cover diverse types of research practices and outcomes, according to the report, instead of "overly-simplified one-stop shops". Here, the MLE confirmed the concerns and recommendations put forward by a dedicated Expert Group on Altmetrics and endorsed the coming activities of a European Forum for Next Generation Metrics.

MLE participants further called for clear goals and missions against which Open Science should be evaluated. Based on cross-national exchanges in the use of altmetrics in policy, the report called for more research on how they could be used not only to promote openness, but also as tools for more profound change – diversifying innovation landscapes and raising awareness of niche pockets of excellence. Altmetrics could also provide visible links between education and science, and help to overcome the problem of research fragmentation across Europe and beyond.

"Participation in the MLE provided a great opportunity to get closer and deeper insight into the implementation of various practices of Open Science. The established contacts and information provided encouraged me to propose concrete measures to our leaders."

Aušra Gribauskiene, Chief Officer of the Science Division of the Ministry of Education and Science of the Republic of Lithuania

It is extremely difficult for researchers to adopt Open Science practices without a broad institutional shift in support and evaluation structures governing their work. Discussions during the MLE revealed that very few Open Science **incentives and rewards** are currently being implemented in participating countries. MLE participants underlined the necessity to develop incentives for different stakeholders: researchers, research organizations and funders, national governments and policymakers.

Since incentives for researchers need to include radical shifts in hiring and promotion procedures, a very good blueprint for future approaches is the Open Science Career Assessment Matrix (OSCAM). This scheme details the different ways that researchers' less visible work and other types of research outputs can be acknowledged or measured.

Given the highly international nature of research networks, international coordination is crucial to the effective implementation of comparable measures. Each country, research funder and research-performing organisation needs to review the extent to which specific incentives will work on the ground, and adapt the requirements discussed in the final MLE report accordingly. MLE participants strongly advocated the further development of EU strategies and policies fostering systemic change in the scientific reward system, including pilot programmes and new instruments for human resources, skills and training.

Where next? A roadmap for Open Science

With diverse positions and national initiatives for Open Science at play, the MLE clearly reflected the importance of modular approaches based on monitoring and regular stakeholder exchange. A model roadmap and recommendations for implementing Open Science is described in detail in the MLE report.

However, in order to trigger systemic change in research and research policy, and to make countries fit for the next EU framework funding programme Horizon Europe, several considerations apply:

- The implementation of Open Science needs to be part of the bigger picture, with discussion on the roles and functions of science in society right now, and an agenda and mission for science and innovation based on openness.

- National strategies for the implementation of Open Science are essential to better understand and align the links between Open Science policies and general STI policies. ERA should be the central platform for the development of national OS strategies.

- Champions and role models are needed to foster the uptake of Open Science practices and create a sustainable transition towards more openness.

- Open Science is enhancing knowledge markets and improving innovation. The synergies of scholarly commons (open-access digital repositories) and the commercial exploitation of research outputs require a systematic review and substantial evidence.

Follow-up activities include many presentations of the MLE – nationally and internationally – broad online and offline discussions of the outcomes, and several dedicated events (e.g. presentations in OS-related committees and meetings), as well as a broader dissemination event in Brussels in November 2018. Experts and country delegates alike will ensure the wide dissemination and discussion of the MLE outcomes and thus contribute to European leadership in Open Science in all that it represents.

For further information:
The Final Report of the PSF Mutual Learning Exercise on Open Science: Altmetrics and Rewards
https://rio.jrc.ec.europa.eu/en/library/mle-open-science-altmetrics-and-rewards-final-report
The   PSF   Mutual   Learning   Exercise   Open   Science: Altmetrics   and   Rewards
https://rio.jrc.ec.europa.eu/en/policy-support-facility/mle-open-science-altmetrics-and-rewards

> *Thirteen countries participated in the MLE: Armenia, Austria, Belgium, Bulgaria, Croatia, France, Latvia, Lithuania, Moldova, Portugal, Slovenia, Sweden and Switzerland. Over the course of one year, the participants met to explore the best ways to tackle the challenges identified, trigger change and optimise the design and implementation of Open Science policy instruments. Several country visits provided the opportunity to learn from hands-on experience.*

7.6   Open Science, the Next Level

In the EU the action plan on Open Science, next to Open Access and Open Data, has now been directed to a series of Missions in which multidisciplinary teams will take on research on themes which have been defined in deliberation with the public, policy makers and private parties. The research aims at the broader fields defined by the UN Sustainable Development Goals (SDGs) and derived concrete issues in science and society. In that respect it seems that in HORIZON EUROPE RRI will meet Open Science in a sphere of deliberative democracy and value driven research.

https://ec.europa.eu/info/horizon-europe-next-research-and-innovation-framework-programme/missions-horizon-europe_en

To boost the transition to Open Access, CoalitionS an international consortium of funders, including the Wellcome Trust, the Bill and Melissa Gates Foundation and Science Europe, supported by the EU and ERC started in September 2018 PlanS. In January 2018, Robert Jan Smits, who worked closely with Carlos Moedas, after nearly eight years stepped down from his position as Director-General of DG Research and Innovation to become the figurehead of CoalitionS. Open Access publishing and the DORA principles have been promoted by PlanS in a paper that CoalitioS published in September 2018 with a final version in the early months of 2019. The idea of PlanS is very much based on APC's which means that authors and their institutions pay to get articles published. PlanS does not allow for paying extra by authors to make their article open in subscription journals, which as argued before is the way researchers could still publish in top tier journals (Nature, Science and Cell for instance) that are in principle not open. PlanS must be regarded as transformatory, aiming in the longer run for true open access journals and platforms which are owned by academia and/or funders and are not commercially or privately managed. PlanS was met with criticism from some scientists who wanted freedom to publish, and as anticipated from the publishers but also from scientists from the

Global South and from institutes and countries where research funding is also hard to get. As researchers in less wealthy countries can neither afford subscription nor APCs, major inequality in science results from APC's and we must consider how to move beyond APC's. PlanS unfortunately still is very much a European Consortium although major institutes and funders in the USA are part of it. It is working to change this rapidly in order to be able to induce the required change in scholarly publishing at a global scale. Therefore, at least China and the USA, but also partners in Africa, South America and South-East Asia must be persuaded that PlanS will also be in line with their needs and cultural values. I refer to a recent publication edited by Martin Paul Eve and Jonathan Gray that provides insightful analyses of the dynamics of the scholarly publishing system with emphasis on the problems of inclusivity and inequality that I here touched up on only briefly.(Eve & Gray, 2020).

In many countries around the world the Open Science movement is gaining momentum. Open Science is boosted right now, since at the time of writing the practices of Open Science daily show their value in the fight against COVID-19. In many countries there are encouraging initiatives and interventions ongoing, but I realize how lucky we are that The Netherlands wants to be a front runner with since 2017 a National Open Science Platform, with a national open science coordinator, Karel Luyben who is also the chair of the EOSC and with the GO FAIR group at Leiden. Moreover, we have a recently launched nationwide program to change the Incentive and Reward System in academia (VSNU, 2019) and a newly designed Strategy Evaluation Protocol (SEP) (VSNU, 2020) for all research in the country, both which are taking the practices and goals of Open Science and a corresponding Recognition and Rewards model fully into account. This is a powerful sign that academic leadership together with the ministry joined forces. In Utrecht in 2018 an ambitious comprehensive Open Science Programme was launched integrating the four major themes Open Access, FAIR Data and Software, Public Engagement and Recognition and Rewards. Next to writing position papers and designing infographics the teams are engaged in bringing the activities with the Board and the Deans to the faculties but also to the different support services of the universities, like Communication and Marketing, HRM, the Library, Student, Research and Education Services and Information and Technology Service. This university wide implementation is a logical component of the 2020–2025 UU Strategy with a choice for Open, Sharing Knowledge, and Shaping Society (www.uu.nl/en/research/open-science.nl). Bottom up, we have seen very interesting and reassuring movements of early career scientists that started Open Science Communities in almost all Dutch universities. Reassuring because it shows that Open Science has reached 'the trenches' where the scientists are in their daily practice but often do not see much change yet.

It Is All About Strategy

In the first weeks of January 2019 something happened to us. We, that is the five members of a committee that had been given the task to revise the Standard Evaluation Protocol (SEP). We met in November 2018 for the first time, two physicists, a social scientist, a historian/philosopher and a

(continued)

biochemist. The SEP is a national research evaluation protocol that is agreed upon by the federation of universities (VSNU), The Royal Society (KNAW) and the Dutch research Council (NWO). The first SEP was in 2003 in use. The protocol is revised every six years. Before the use of a SEP the University federation had a national protocol to evaluate whole disciplines. Interestingly, and luckily I would add, based on the numerical scores there was reputational competition but not (re)allocation of research funds or university lump sum funding (van Drooge et al., 2013). At the start of our committee work we were told that the feeling was that the SEP 2015–2021 was satisfactory and that only minimal changes were required. We realized however that new research evaluation protocols had been proposed in 2013 by KNAW committees for engineering, the social sciences and a national protocol for the humanities. At the same time, a consortium under the name of Quality and Relevance in the Humanities (QRIH) had produced a protocol for the humanities, which in 2019 was a few years in use (https://www.qrih.nl). It was no coincidence that one of us, Frank van Vree, had been prominently involved in QRIH as Dean of the Faculty of Humanities of the University of Amsterdam. Finally, there was a national debate on incentives and rewards going on.

We in November and December 2018 in two meetings discussed the previous SEP which at first sight indeed looked very good (VSNU, 2014). It was written in 2014 with amendments added in 2016. Compared to its predecessor it had downplayed the emphasis on quantity, productivity, metrics and thus the aspect of national competition. It stressed relevance to society and, a bit to our surprise I believe, made it clear that *'the research unit's own strategy and targets are guiding principles when designing the evaluation process.'* (p5) The working group that in the background supported us had obtained an evaluation report on how SEP had been actually used by research evaluations in the recent past. There was limited data, but it left the impression that the intentions and prescriptions of the SEP had not been followed. With respect to huge differences in research practices and academic output, the degrees of freedom offered by SEP had also not been taken advantage of. Audits still very much were focused on quantitative output (papers, JIF and h-index, books published by specific publishers) and research grants won.

These first days of the new year it dawned on us that the SEP was not the problem, it could be easily updated with new developments like DORA, Open Access, Open Data and the other aspects of Open Science that are less well known. The problem was the way the evaluations were done and how poorly that connected to the context of the researchers, their research and to our relationship with society. The research evaluations were experienced as a heavy burden, with little noticeable effect and not thought of as an interesting opportunity for reflection on strategy and goals looking for improvement in discussion with colleague's, peers but also with Deans and the Board of the institute. We decided to take time to think this through and organized in February and

(continued)

March several combined lively meetings with the working group who responded enthusiastically to this intervention. We made a choice to not only assess 'the what', the quality and diverse impacts of research results, which is more or less the usual assessment. We wanted in the new SEP to emphasize the evaluation of 'the how'. How is a research unit managed and organized, is there a deliberate strategy for research, but also with respect to leadership, HRM, integrity, safety and diversity? How is the unit connected to scientists in other disciplines and to stakeholders in society? Is there awareness of relevant developments in science and the world? This turn was generally understood and well received. Working towards a draft version of the new SEP after the summer, we discussed in small and larger national meetings the new items like DORA, Open Science, the use of narratives and numerical scores and the idea of *academic culture* with deans, rectors, directors of institutes, researchers and university policy advisors.

There were issues and worries. The idea that researchers in departments and research centres should have a research strategy beyond production of papers and winning grants which would be the start of the evaluation process was not always immediately accepted. It was felt to be problematic that the evaluation looked to the strategy of a unit and thus was incomparable between units doing research in the same discipline in another universities with different strategic aims. This and abandoning the use of numerical, 'absolute', scores was felt to introduce subjectivity since it made comparisons within and over disciplines impossible. We argued that this sense of objectivity in comparing apples with pears was anyhow false to begin with. There was a feeling that narratives of researchers, which were proposed to explain strategic aims, plans and results might be used to cover up weaknesses by smooth and slick language. The narratives of the audit committee, it was expected, would be vague, non-critical and useless. Some, as expected, suggested: *'Wouldn't journal metrics and a final score of 1 to 5 be more objective and thus better? It takes also much less time than reading and discussing the science.'* We listened during the year carefully to these very diverse opinions, worries and comments which we used to improve the SEP until its final version of December 2019 (VSNU, 2020). We knew that for the use of this protocol with a new, more meaningful, way of reflecting on and assessment of research, researchers and policy makers will need help from experts. We realized as I conveyed in this book, that science is in transition, more than we had anticipated a year before and that therefore the gradual change of the SEP, the research evaluation indirectly linked to the academic incentive and rewards system, was logical if not inevitable. Finally, but importantly it was decided and accepted to change the name of Standard Evaluation Protocol to the more appropriate Strategy Evaluation Protocol.

With these developments with respect to our thinking about the way we should do research evaluations, in our country and also abroad, I would say clear progress has been made. We are in transition to begin using evaluation schemes that recognize and respect that science and scholarship in its goals and practices are in essence pluriform, must be open, inclusive and diverse and should allow for the 'outside-in' perspective by those who are stakeholders to our research in the wider society. It puts emphasis next to its products also very much on the process and practice of research. Finally, and most importantly the evaluation has to be performed integrally from the perspective of the aims and strategy of the research unit that is being evaluated. Strategy and aims may be confined to the domain of science and knowledge for its own sake but may very well also be inspired by societal challenges, regional, national or international. This changes the credit cycle as shown below to be truly open to and collaborating and sharing with relevant agents in society.

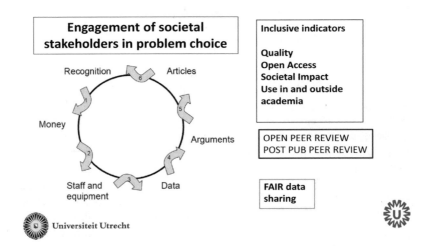

7.7   The Sceptics: 'Open Science is either naïve or the next neoliberal trick'

The unwanted influence and distorting effects of political, economic or other forms of abuse of power was discussed in relation to Dewey's deliberative democracy and Rawls' well-ordered society, Kitcher's well-ordered science, the Public Engagement movements and Mode-2 science in Chap. 4 and 5 (see for the critique for instance (Halffman & Radder, 2015) (Fuller, 2000). Some have argued that Open Science is either naïve or the next neoliberal trick. There are those who have a deep concern, if not a total distrust, regarding the pervasive influences of late capitalism and its neoliberalism in current politics in almost all countries around the world. The same worries have been expressed and apply in the context of Open Science (Mirowski, 2018). There are two main and very different responses to the worries about public engagement in a 'post-truth' age, which resonate the Dewey-Lipmann debate one hundred years ago discussed in previous Chapters. The first response is Open Science, Deweyan, as is well-ordered science and in a sense Mode-2. The other reaction is returning to a more classical insulated way of doing science, distrusting external influences and protecting science and research from these influences and in essence going back to Mode-1. I have argued that this response does not seem to take into account the major changes in society, which are undeniably of great relevance for the practice of science and its relationship with the citizens in society and its role in policy making in democracies (Habermas, 1970, 1971; Jasanoff & Simmet, 2017). Jasanoff and Simmet correctly stated that 'post-truth' and 'alternative facts' are powered by, but not caused by the internet or social media, and have existed in different forms in any human civilization. Science and experts have to deal with that by engaging and debating. Romanticising the authority of the classical myth of science and scientism is not the way to go.

With regard to Open data and Open Access it has been noted correctly that in these movements the perspective is mainly that of the richer countries. I refer to a book on the different perspectives and worries about Open Access in *Science and the politics of openness (Nerlich et al., 2018)*. including a chapter by Stephen Curry who is the leading person of DORA. I already referred to the book edited by Eve and Gray elaborating on these issues (Eve & Gray, 2020).

Subscription fees, especially in combination with Article Processing Costs (APCs) the latter now central in most OA models, are beyond financial capacities of researchers in large parts of the world – for instance South-East Europe, Africa, South America, Indonesia and India. Will the richer more powerful countries benefit more from Open Data? From that perspective as a reflex, some argue science must be insulated and not be made more open and not be rendered more vulnerable to these external powers than it already is. As I have argued above, based on recent history, that is not the way science should develop, given the socio-economic and public health issues we are facing now and will be facing in the near future.

Open Science, Open Eyes to the World

Isn't it obvious that implementing DORA, which forbids the use of JIF as a proxy for quality is a blessing to all those who were for the wrong reasons left behind? Isn't making journal articles Open Access, by having the authors pay Article Processing Costs (APCs) an important step? Now everybody, everywhere can read without costs. Don't we all agree?

(continued)

Prof. Mamokgethi Phakeng Photo courtesy of Lerato Maduna / University of Cape Town

Prof. Mamokgethi Phakeng and the author, Prof. Frank Miedema at the UCT conference, Cape Town, South Africa, June 2018. Photo taken by Carolyn Newton

Not that afternoon, December 2019. I was in KU Leuven and took part in a discussion on: 'Open access in a global perspective: comparing policies and practices'. Three expert speakers presented their views of the Open Access movement, including PlanS. From their perspective from the East and the South (Mexico, South Africa and Indonesia), Plan S was no good. Subscription fees but also APCs, ranging from 550–5000 USD per article, are in their part of the world far beyond financial reach of scientists. Even worse, at the same time they are required by their institutions, who want to climb the Rankings, to publish their work in the high impact journals. Moreover, regarding publishing Open Data, because of the minimal research options and budgets, despite their good ideas, they would not be able to profit from Open Data. But others with more budget will be able to use their data, they said. One of the speakers shared with us his deep-felt worry that in his country where half of the population lives in poverty, universities demand papers in high ranked internal journals which forces researchers away from badly needed research on local and national problems.

(continued)

After these for us confrontational talks, in the debate session I could only humbly admit that the speakers were totally right and fully entitled to put this critique to us, scientists from rich countries. I realised, we have these experiences, but we must work much harder to reflect on them and have them influence our way of doing and improving science. I referred to my learning experience in a multidisciplinary ten-year project on HIV/AIDS in Ethiopia. In the Ethiopian Netherlands AIDS Research Project (ENARP), the Amsterdam HIV/AIDS researchers in close collaboration with The Ethiopian Health and Nutrition Research Institute in Addis Ababa between 1993 and 2003, had set up a large prospective study on HIV and AIDS. In that very productive international project, we were continuously reminded by our Ethiopian colleagues about their vision about how the project should be executed and about research on the local needs of the public. HIV/AIDS was in Ethiopia a very different disease which for instance affected women and children much more then in Europe. Doing scientific research also was a different social endeavour compared to Amsterdam because of the local socio-economic situation and of course totally different cultural values and beliefs. At times we were reminded that because we had more money, we should not think that we were smarter and better scientists. In addition, I mentioned my more recent experience at a meeting at Cape Town University in June 2018 with the topic: *Beyond the Metrics: Measuring the Impact of Research*. There young researchers working in the Townships showed us what science has to offer when addressing the needs of local populations struggling with poverty and disease. This was preceded by an impressive introduction by Professor Mamokgethi Phakeng the Vice-Chancellor of UCT about science and democracy which I here reproduce in its entirety.

7.8 Beyond the Metrics: Why Care About How We Assess Research Impact?

Prof. Mamokgethi Phakeng, June 26, 2018.

UCT has been grappling with the challenges of how to measure the impact of science beyond bibliometrics – in particular, the effect of the current system on the progress of young researchers and on encouraging socially responsive, interdisciplinary research that addresses South Africa's and the continent's problems. At a recent symposium intended to bring together thought-leaders to challenge our thinking and help us start to develop alternative tools, Professor Mamokgethi Phakeng introduced the event during the week before she assumed her position as UCT's vice-chancellor. What follows are her thought-provoking words.

Why do you do the research that you do? What has been the impact of your research so far? What has it informed? What has it changed? What difference has it made?

How different would our society be without it? If you hadn't done the research that you have done so far, what is it that we would have missed? Most importantly, how do you know that your research is making an impact? What is it that tells you that your research has had impact?

These questions are as relevant to individual researchers as they are to university management, funders and government, and this is the reason why we need to consider the important issue of moving beyond metrics to measure research impact.

Measuring the impact of our research is about considering what happens as a result of our research. That's the tricky part, because it means a researcher can't actually "do impact". You can only undertake activities that enable impact to happen. The questions we ask, the theories, approaches and methodologies we use, as well as how we do our analysis and present our findings are all important and can enable impact. And of course, impact will happen when others take-up and use our research to change something.

I want to offer three provocations with a hope that they filter into our conversations about this important topic.

What are the shared values between sound research and sound democracy?

The first provocation is that research impact is a result of not only the knowledge that is produced. We should look at the impact of research also from the perspective of the values that the practice and process of research inculcates, especially in a young democracy such as ours.

But does research have a place in building democracy?

There are important similarities between research and democracy: the sound conduct of research and the sound conduct of democracy both depend on the same shared values. The very virtues that make democracy work are also those that make research work: a commitment to reason and transparency, an openness to critical scrutiny, a scepticism towards claims that too neatly support reigning values, a willingness to listen to countervailing opinions, a readiness to admit uncertainty and ignorance, and a respect for evidence gathered according to the sanctioned best practices of the moment.

Looking at research impact from this perspective not only elevates research, but it also elevates democracy. Of course, we can argue about whether research has a unique claim on these shared values. That is not important at this stage: what is important is that these values are critical, especially in our country where we must build a culture of democracy.

In strengthening democratic values, we also renew the preconditions for scientific discovery and technological innovation, and thus, high-impact research. The converse of this is also true – research with impact can serve as a precondition for building a vibrant democracy.

Should all of our research be for the public good?

The second provocation is that questions about research impact are often about the contestation of resources: where we invest them and why. We spend billions in public money on research annually, so we have to be accountable and consider its social, economic and environmental impact, as well as its impact on health and well-being and technological developments.

Questions about research impact also force us to consider whether our research spending is the best way to use our very limited resources. Considering the source of our research funding, it is very easy to argue that research should be for public good. But the question is, should all of our research be for the good of the public?

My view is that it is important to have a balance between curiosity-driven and mission-oriented research; research that tackles fundamental questions and research that serves corporate interests; and research for pleasure and research for pay.

Despite the need to engage in research for the public good, it is also necessary to create safe spaces where smart minds can tackle hard questions without any expectation of immediate applications. Like democracy, research is also a value to pursue for its own sake. The argument for engaging in curiosity-driven research that tackles fundamental questions or research for pleasure is always going to be difficult to sell because in a developing country, such as ours, research has a big responsibility to respond to the triple-challenge of unemployment, inequality and poverty. But scientific knowledge is a public good, therefore courageous policymakers and funders should be prepared to pay for that resource without imposing a utilitarian approach on all publicly funded acquisition of knowledge.

How does the way we measure impact shape our research?

The third provocation is that the way we measure impact has implications for how we recognise and reward performance. It will therefore shape our research activity, research output and research training. If metrics drive research, the danger is that research can become formulaic – focused only on getting citations and impact factors right for purposes of career advancement and winning grants. This will encourage unethical behaviour and destroy our scholarship. We can already see this in the increase in the number of predatory journals as well as people who publish in them. This is why we need to be clear about what counts as research impact and how we measure it.

https://www.news.uct.ac.za/news/research-office/-article/2018-07-05-why-care-about-how-we-assess-research-impact

7.9 Open Science in an Open Society

These problems and threats to science are very real as we do witness these days both in democracies and less democratic state-capitalist systems. Obviously, from the economic perspective, Research and Innovation is a main driver of economic growth and job creation. This is clearly stated in most national, EU or international plans about science and technology and has been the dominant driver in the recent past. At the same, because of hard work and lobby, I am sure, societal targets and societal impact have a firm place in the agenda and the social sciences and humanities are increasingly building their case in these times of the Covid-19 pandemic. SDG's and Grand Challenges are inclusive and perceptive to social needs and values. As argued in the previous chapters, in a true pragmatist vision, research and its subsequent social actions must be inclusive and continuously reach beyond classical technocratic scientism. In this book I have focussed on the role of science and research and how it must be organized and reshaped to contribute more and differently. It is clear that Open Science, which engages in a truly open relationship with the public can only optimally contribute to an Open Society with a certain minimal degree of democracy. It is not for science to decide how politics and the public sphere is organized and regulated. Given the above, however, the engagement of scientists, irrespective of their political views, to contribute as public intellectuals in this debate, and thus in political discussions about institutions and the proper functioning of democracy is required.

Finally, one may wonder, will there ever be One Science in the sense of a world-wide truly Open Science Practice? That was the belief of the previous generations after WWI and WWII, but we have seen how this, despite modernization and globalization, has developed in our present hyper-modern times. It seems we are not even near, but my hope is that Open Science may well be one of the best instruments to align science under a global banner and Europe must be optimistic and lead the way as it successfully did in the COVID-19 crisis at the time I wrote these lines.

The New York Times, May 4, 2020

World Leaders Join to Pledge $8 Billion for Vaccine as U.S. Goes It Alone

The E.U. organized a teleconference to raise money for coronavirus vaccine research, drawing contributions from around the world.

BRUSSELS – Prime ministers, a king, a prince and Madonna all chipped in to an $8 billion pot to fund a coronavirus vaccine.

President Trump skipped the chance to contribute, with officials in his administration noting that the United States is pouring billions of dollars into its own research efforts.

A fund-raising conference on Monday organized by the European Union brought pledges from countries around the world – from Japan to Canada, Australia to Norway – to fund laboratories that have promising leads in developing and producing a vaccine.

(continued)

For more than three hours, one by one, global leaders said a few words over video link and offered their nations' contribution, small or large, whatever they could muster. For Romania, it was $200,000. For Canada, $850 million. It was a rare show of global leadership on the part of the Europeans, and a late-hour attempt at international coordination. Countries the world over have been pursuing divergent – and often competing – approaches to tackling the pandemic.

References

Bartling, S. (Ed.). (2014). *Open Science: The evolving guide on how the internet is changing research, collaboration and scholarly publishing*. SpringerLink.

Beck, U., Giddens, A., & Lash, S. (1994). *Reflexive modernization : Politics, tradition and aesthetics in the modern social order*. Polity Press.

Burgelman, J.-C., Pascu, C., Szkuta, K., Von Schomberg, R., Karalopoulos, A., Repanas, K., & Schouppe, M. (2019). Open Science, open data, and open scholarship: European policies to make science fit for the twenty-first century. *Frontiers in Big Data, 2*. https://doi.org/10.3389/fdata.2019.00043

EBRARY. Restricted to UCB IP addresses. http://ebookcentral.proquest.com/lib/berkeley-ebooks/detail.action?docID=3339454.

ESF. (2013). Science in society: Caring for our futures in turbulent times. Retrieved http://archives.esf.org/uploads/media/spb50_ScienceInSociety.pdf

EU. (2016). Open innovation, Open Science, open to the world. Retrieved from https://op.europa.eu/s/sTsE

EU. (2017a). Evaluation of research careers fully acknowledging Open Science activities. Retrieved from https://orbi.uliege.be/bitstream/2268/215460/1/os_rewards_wgreport_final.pdf

EU. (2017b). Next-generation metrics: Responsible metrics and evaluation for Open Science. Retrieved https://op.europa.eu/en/publication-detail/-/publication/b858d952-0a19-11e7-8a35-01aa75ed71a1

EU. (2018). Open Science policy platform recommendations. Retrieved from https://www.go-fair.org/2018/07/02/2351/.

EU. (2019). Indicator frameworks for fostering open knowledge practices in science and scholarship. Retrieved https://op.europa.eu/en/publication-detail/-/publication/b69944d4-01f3-11ea-8c1f-01aa75ed71a1/language-en/format-PDF/source-108756824

Eve, M. P., & Gray, J. (2020). *Reassembling scholarly communications: Histories, infrastructures, and global politics of open access* (pp. 1 online resource (xxvii, 438 pages)). Retrieved from https://ieeexplore.ieee.org/servlet/opac?bknumber=9255850 IEEE Xplore.

Fecher, B., & Friesike, S. (2014). Open Science: One term, five schools of thought. In F. S. Bartling (Ed.), *Opening science*. Springer.

Fuller, S. (2000). *The governance of science : Ideology and the future of the open society*. Open University Press.

Habermas, J. (1970). *Toward a rational society*. Heinemann Educational Books.

Habermas, J. (1971). *Knowledge and human interests*. Beacon Press.

Halffman, W., & Radder, H. (2015). The academic manifesto: From an occupied to a public university. *Minerva, 53*(2), 165–187. https://doi.org/10.1007/s11024-015-9270-9

Jasanoff, S., & Simmet, H. R. (2017). No funeral bells: Public reason in a 'post-truth' age. *Social Studies of Science, 47*(5), 751–770. https://doi.org/10.1177/0306312717731936

Impact factor abandoned by Dutch University in hiring and promotions, Nature June 25, 2021.

Mirowski, P. (2018). The future(s) of Open Science. *Social Studies of Science, 48*(2), 171–203. https://doi.org/10.1177/0306312718772086

MIT Press Direct. (If available from multiple sources, select The MIT Press Direct). Restricted to UC campuses https://doi.org/10.7551/mitpress/9286.001.0001.

Nerlich, B., Hartley, S., Raman, S., & Smith, A. T. T. (2018). *Science and the politics of openness: Here be monsters*. Manchester University Press.

Nielsen, M. A. (2012). *Reinventing discovery: The new era of networked science*. Princeton University Press.

Owen, R., Macnaghten, P., & Stilgoe, J. (2012). Responsible research and innovation: From science in society to science for society, with society. *Science and Public Policy, 39*(6), 751–760. https://doi.org/10.1093/scipol/scs093

Owen, R., Bessant, J. R., & Heintz, M. (2013). *Responsible innovation: Managing the responsible emergence of science and innovation in society*. Wiley-Blackwell.

Project MUSE. https://muse.jhu.edu/book/46989.

Project Muse. Restricted to UCB IP addresses. https://muse.jhu.edu/book/46989/.

Rentier, B. (2019). *Open Science, the challenge of transparancy*. Académie Royale de Belgique.

Stilgoe, J., Owen, R., & Macnaghten, P. (2013). Developing a framework for responsible innovation. *Research Policy, 42*(9), 1568–1580.

Suber, P. (2012). *Open access* [text]*MIT Press essential knowledge* (pp. 1 online resource (xii, 242 pages)). Retrieved from MIT Press via IEEE Xplore. Restricted to UCB, UCD, UCI, UCLA, UCM, UCR, UCSC, and UCSD http://ieeexplore.ieee.org/servlet/opac?bknumber=6267549

Suber, P. (2016). *Knowledge unbound selected writings on open access, 2002–2011*. MIT Press.

van Drooge, L., de Jong, Stefan, Faber, Marike, Westerheijden, Donald F. (2013). Twenty years of research evaluation. Retrieved https://research.utwente.nl/en/publications/twenty-years-of-research-evaluation

Von Schomberg, R. (2011). Prospects for technology assessment in a frame of ressponsible research and innovation. In R. Dusseldorp & M. Beecroft (Eds.), *Technikfolgen abschätzen lehren:Bildungspotenziale transdisziplinärer Methoden*. Wiesbaden Verlag.

von Schomberg, R. (2019). Why responsible innovation. In R. von Schomberg (Ed.), *The international handbook on responsible innovation. A global resource* (pp. 12–32). Edward Elgar Publishing.

von Schomberg, R., & Hankins, J. (Eds.). (2019). *International handbook on responsible innovation : A global resource*. Edward Elgar, Publishing.

VSNU. (2014). Standard evaluation protocol 2015–2021. Retrieved from https://www.knaw.nl/nl/actueel/publicaties/standard-evaluation-protocol-2015-2021

VSNU. (2019). *Room for everyone's talent., towards a new balance in the recognition and rewards of academics*. Retrieved from https://vsnu.nl/recognitionandrewards/wp-content/uploads/2019/11/Position-paper-Room-for-everyone%E2%80%99s-talent.pdf

VSNU. (2020). Strategy evaluation protocol 2021–2027.Retrieved from https://www.vsnu.nl/files/documenten/Domeinen/Onderzoek/SEP_2021-2027.pdf

Chapter 8
Epilogue: Open Science in an Open Society

Abstract The European Union has chosen Open Science as the way to do science and research based on its cultural and social values. Open Science can only really thrive in democracies and Open Societies to the benefit of humanity. This relationship between science, scientists and society is not trival and sometimes endangered, therefore we need to continuously engage in research with and for society.

I have described in detail the changes in science and more briefly the changes in society since 1945 that are of greatest relevance to the current practice of science and research. The old way of thinking about and doing science does not fit with the dynamics and needs of social life and society in this age of hyper-modernity. I have discussed in depth why and how research and academia have to change to make both of them fit for the future.

I have followed experts, philosophers, sociologist, historians and STS scholars in the evolution of their thinking about science. I have shown that since the 1980s they have gone back to old, but realistic concepts of pragmatism of how science produces reliable knowledge and how it will increase its impact. These concepts are still valid and since 1980 have been revitalised and modernized by the most influential and visionary thinkers about science in society. This intellectual journey eventually, and I argued in many respects inevitably, led us to Open Science, an inclusive deliberative and democratic way to set the agenda of science and research in connection with 'the publics and their problems' in society. By doing research according to the practice of Open Science we will in a truly inclusive way, appreciate and accommodate all kinds of academic research and their different excellences and products. I have argued that this is required to successfully take on the Sustainable Development Goals (SDGs) and the Grand Societal Challenges of our time having said this, it must be realized that Open Science needs an Open Society, and vice versa the Open Society needs Open Science to contribute in a balanced way to its social and economic future.

That future has already started and is badly in need of knowledge produced by science performed according to the model of Open Science. Writing these lines, we are daily witnessing how the COVID-19 pandemic since March 2020 has devasting effects on global health and prosperity and hence our societies and the lives we live all over the world. In the final lines of the previous chapter on Open Science I asked the question: 'Will there ever be One Open Science in the sense of a worldwide community committed to practice Open Science?' In the midst of the COVID-19 pandemic, the relationship between science and society becomes immediate and urgent and is shown to be absolutely critical to our future, for the short and for the longer term. When the damage to public health and the personal drama of literally hundreds of thousands of deaths caused by the virus are in the daily news, and the worries about its effects, both on health and economics in the other parts of the world are increasing, experts and their science were immediately on the problem and at the heart of policy advice and policy making. The general insecurity and the broadly felt lack of control caused by an invisible virus directs the attention of the public to the scientific experts and they look up to them with expectation and hope. Virologists and epidemiologists first, and then social scientists and economists were asked to comment and give their views, on how to keep us healthy and on what will happen to the economy and our social life. How come we didn't see it coming? Can we get back our control? Can a next pandemic by such a virus be prevented or at least more rapidly controlled by science?

Fortunately, the scientists were almost all honest brokers, giving honest answers about what science is and what its limitations are. We have to admit, and anticipate, this leaves the public behind with fears of high risk, lack of control and insecurity with respect to their personal future. As expected in major crises, populist parties and other groups in society playing to the emotions of their electorate and followers showed distrust of science and experts in a wave of anti-elitism. Experts and politicians in the meantime try to be as transparent as possible with respect to the incomplete data, that was literally changing and improving day by day, for their analysis and advice, and the seperate process of political decision making, respectively. Apart from some loud minorities who did not believe COVID-19 existed or was harmful at all, the trust in science was high. It is quite disturbing that President Trump and other elected presidents for shorter or longer times seemed to openly sympathize or belong to these minorities. Geopolitics interfered with the free flow of scientific data on the origin of the virus and resulted in the usual blaming and scapegoating. This explains Trump's no-evidence accusations on the part of China and the censoring and silencing of researchers and civilians by the Chinese government.

In the meantime, the international scientific organization of the open exchange of data on the molecular biology, receptor use, sequence of the viral RNA of the virus and specimens and research material and data on the course of the pandemic -prevalence, hospitalization, ICU needs, mortality and morbidity- was a true example of a near global Open Science practice. Despite Trump's believe May 9th, 2020 that the virus would spontaneously go away without a vaccine, subsequently, initiatives to establish (pre)competitive global initiatives to develop therapy and vaccines were launched. Of course, given how we have decided to organize the production of vaccines and medicines in our societies, the interaction with pharmaceutical companies is economically complex. They normally have to please their shareholders while operating in the international markets, but now for COVID-19 were made to commit to affordable prizes based on transparent costs, but patents are allowed. As of this writing a couple of vaccines have already been approved, have shown excellent protection and vaccinations have started and shown success. With a pandemic which appears to be extremely difficult to control, this is a major achievement and hopefully will have a major effect on the course of the pandemic worldwide. The development of COVID-19 tests and treatment modalities, but in particular the historic quick and large-scale development of different vaccines is widely heralded as a major triumph for science, comparable by some to the Manhattan project. The academic publishers have opened up their paywalls to provide open access to articles related to corona viruses and COVID-19. After the applause dies away, it prompts the obvious question why this is not common practice because thinking about biomedicine and health alone, cancer, stroke, cardiovascular disease, asthma, dementia, and Alzheimer's and many other diseases aren't they not also a major threat to our health. Wouldn't this be very helpful for research and innovation on climate change and our thinking how to work on inequality, institutions in open societies and the many other fields of research and scholarship including ethics, political philosophy, research on socioeconomics badly needed to guide our actions in complexities of the real world?

Why can we only mobilize science and scientists and academic publishers in times of intense crises and of war? I agree with Marianne Mazzucato who has with endless energy and high visibility and impact argued for more direction and guidance from governments in democracies to organize our science and development according to large societal missions (Mazzucato, 2013) (Mazzucato, 2018). In the EU, as I discussed in Chap. 7, this has already started in Horizon 2020 and is an even more pronounced founding principle of the Horizon Europe program that will from 2021.

8.1   Open Society

This idea of an Open Society, or rather the lack of it, may be the problem for optimal development and implementation of Open Science in certain regions of the world. In the COVID-19 crisis, because it started unquestionably in mainland China, we were on a near daily basis tutored on Chinese history and politics. This tutoring already began with the trade conflicts between the USA and China and by the increase of Chinese interventions in Hong Kong in 2019. For some reason, we in the EU, with the Obama administration, had high hopes for reforms in China which would we thought bring the country more towards personal freedom and some form of liberal democracy. In the past 15 years many universities started very active collaborations with counterparts in mainland China. Many had such collaborations in Hong Kong and Singapore to keep in contact with the Far East and its huge investments in higher education and research. The Chinese Academy of Science actively started efforts to improve the quality of the research which was being critiqued in recent years in *Nature* and *Science*. In the more recent years however, these expectations of societal and political change did not materialize. On the contrary, the Chinese Communist Party and their leader who holds absolute power, have adopted state capitalism and gone back to their old concepts about politics, the state and society. Their goal clearly is not only to become an economic and political superpower, but also to show that their model of the state and society is superior to the Western liberal or social democracies. In this major scheme, it places interests of the state, determined by the CCP, above personal freedom of its citizens. It is exactly the latter that has, with the Enlightenment, brought Modernity to the West which allowed for the development of modern science. The experts on China tell us how to understand these developments in the context of the past hundred years of Chinese history, and in the present, China's interventions among others in Hong Kong. China is rapidly developing to become a global superpower in science and technology, which for science and open science means that there will for years to come not be one global community, not one way of doing science and research.

There actually never was one global science community, but after 1989 there was a brief moment in time when we believed that it might be possible to have a global science, which we in the West erroneously thought per definition would be our way of doing science. The way Chinese society and its science are governed does not allow for science and research to be performed in open deliberative relationship with the publics and their problems as I have depicted in Chap. 5 as the ideal. This does not mean that there will not be collaboration, exchange and discussion with Chinese researchers. There are many grand and global challenges, such as COVID-19, economics and climate change in which global collaboration in research consortia are to the benefit of us all, no matter our different national political systems. To investigate these problems and their solutions, normative political choices with regards to science have to be made. For this academic leadership at every level of the science system, national and international, funder or academia has to step up to the plate. The EU has chosen Open Science as the way to do science and research. It are the political, cultural and social values of the EU in which we have to keep investing to see to it that Open Science can thrive in democracies and Open Societies to the benefit of all people (Wilsdon & Rijcke, 2019).

Hong Kong, Where the West meets the East.

Utrecht University has in the past seven years invested in an institutional research collaboration with Chinese University of Hong Kong. This has been established by several professors and members of their research groups and by spin outs of UMC Utrecht. I first visited CUHK with a UU life science

(continued)

delegation in December 2013, at which occasion a collaborative agreement between the universities was signed by Marjan Oudeman, the Utrecht University President and the Vice Chancellor Joseph J.Y. Sun. There are many common research interests, from regenerative medicine, 3D printing of cartilage to be surgically applied to the knees of the affected elderly, to large cohort research on schizophrenia and public health in relation to air pollution. Professor Tuan, a very open and dynamic personality, was the leader of the orthopaedic regenerative medicine program. He had an impressive career in medicine in several top-notch universities the USA before he returned to Hong Kong. Opportunities for collaboration were discussed with major universities in mainland China and especially with the enormous new research and biotech facilities that were being build, just across the border, in Shenzhen. We fully realized these opportunities and the commitment of China to science and technology, since we had just arrived from Beijing where we visited Peking University and had meetings over dinner with government officials. I remember the magnificent view overlooking Hong Kong from the high hilltop on which the Board of CUHK has her offices and where we discussed science and research over lunch in a very friendly and open atmosphere.

In January 2019, I returned to CUHK, with an Utrecht University delegation now lead by our President Anton Pijpers. A warm welcome was expressed by professor Rocky Tuan, the CUHK's Vice Chancellor since January 2018. I passionately delivered my short opening talk on Open Science in UMCU and UU and our aim to increase impact by public engagement. I was, I am afraid, a bit hyped-up since because of delays, we came straight from the airport. Besides, that the speaker on behalf of the host started with the usual figures, metrics and the Shanghai ranking of CUHK, the idea of science for real impact was met with sympathy and we heard that similar actions had started in CUHK. We again had very constructive conversations and exchanges and, though we did not visit CUHK at Shenzen, were impressed by what had there been achieved. This clearly was a big opportunity with indeed major investments coming also to the advantage of CUHK. In the evening we took part in a meeting with UU students and alumni at the Dutch Consulate. The Consul reassured us, in a private talk, that Hong Kong was politically stable, and that Beijing was more than happy with Hong Kong as an international business and science hub. The recent unrests were to be regarded as minor incidents, no need to worry about.

Things have taken another course though. It happened that I was again in Hong Kong in the first week of June 2019, to give a talk at the 6th World Congress on Research Integrity (https://wcrif.org/wcri2019). The meeting was hosted by Hong Kong University in the person of Mai Har Sham, **Associate Vice-President (Research),** a biomedical researcher by training who has a strong track records in research integrity. It was a very international meeting with a program that touched upon the different levels of academic integrity, including students, PhD's, professors but also at the institutional Board level of Deans and Vice Chancellors. As discussed in Chap. 7, I discussed Open Science as a practice of science and research which may help to improve research integrity at all levels. I stressed that for this to be successfully achieved, we have to reflect on our way of doing research, how we organize academia, and about our ideas about the relationship with society. This resonated well with the opening statement of the congress, which emphasized the need to reflect on cultural differences and how they influence our ideas about science. There were plenary presentations by colleagues from universities and government agencies around the world and from the Chinese Academy of Sciences. In her presentation professor Mai Har Sham touched upon the current issues regarding scientific integrity and the actions that were ongoing.

(continued)

After a presentation by a colleague from a Chinese university about the university's actions and code of conduct, critical questions from the floor regarding problems with integrity and intellectual autonomy caused inconvenience by the speaker and the audience. Here we witnessed that Open Science needs to take into account, reflect on and continuously discuss cultural and political differences, that are deep seeded in society and ingrained in its practice of science. That evening, when the first of many marches against the influences of the Chinese government on Hong Kong democracy took to the streets, at the congress dinner party I shared a table with young successful civil servants from Beijing confining ourselves to small talk about science, their lives in Beijing, kids and parents.

Taking stock of science in the COVID-19 crises, it seems that science and scientists as an international community are committed and more than ready to practice Open Science. However, the open society- with its plurality, economic inequality, the speed and the use and abuse of social media, the higher levels of education, but also the increasing differences in education levels, the populism fuelled by politicians- is often felt to make the connection between science and the public no less complex and to someeven dangerous. Social media and the role of the tech giants since 1990 have had an enormous impact on how, when and where the debates in the public sphere take place. Fuelled by ugly partisan battles, the internet it seems has divided countries and people more than it has resulted in open debates, in which listening to each other's fears and opinions is being practiced, to reach mutual agreements. This is a major problem for science and society. Recently we have seen the worst of it in the USA, where partisan battle lines already since the 1980s are raging. Despite the ideals of the Founding Fathers and the Constitution, before and after the Progressive Era of 1890-1920 or FDR's New Deal, the USA has seen such ugly episodes before Google, Facebook, Twitter, the internet and cable news with CNN, Fox News nearly wiped out serious media and national newspapers. These episodes have to a great extend determined politics in general and the politics of science in particular (Diggins, 1992, 1994). Lepore's impressive history of the USA, through the lens of the Declaration of Independence (1776) and the Constitution (1787), is a surprisingly gloomy reading experience (Lepore, 2018). The Founding Fathers clearly anticipated the ugly episodes with partisan battle lines, so we must, nor in politics nor in science be naïve, but we must from academia engage in continuous debate with policy makers and the various publics. There are many experiences showing that engaging in serious discourse about contrasting ideas and convictions is helpful to reach levels of understanding, if not common ground about issues in social life. Moreover, as Habermas argues, these deliberations, more than our voting, make our society truly republican and democratic and 'we must find knowledge through these deliberations and utterances in the social context were the action is' (Diggins, 1994)(p365).

The time is long gone that the claims and views of science and experts were automatically accepted because of mythical 'God given' authority or a 'unique scientific method'. As I have argued and demonstrated, the sciences, in their many different communities of inquirers do produce reliable and robust knowledge that has proven successful and has in the past contributed enormously to the quality of life. Much is still to be done and at this very moment scientist around the world are working 24/7 on therapies and vaccines for COVID-19 which are badly needed. To make clear what science has to offer we have to engage tirelessly in continuous conversation, debate and discussions about science and society. With the same energy and perseverance, because of geopolitics, ugly partisan politics and outright suppression we have to keep campaigning for open debates and deliberative democracies, as the stakes for humanity are higher than ever, this needs to be done within our own region, country, in the EU and in global collaborations around the globe.

References

Diggins, J. P. (1992). *The rise and fall of the American left*. W.W. Norton.
Diggins, J. P. (1994). The promise of pragmatism: Modernism and the crisis of knowledge and authority. In *Chicago*. University of Chicago Press.
Lepore, J. (2018). *These truths: A history of the United States* (1st ed.). W.W. Norton & Company.
Mazzucato, M. (2013). *The Entrepreneurial State Debunking Public vs. Private Sector MythsAnthem Other Canon Economics* (2nd ed., pp. 1 online resource (284 p.)). Retrieved from http://kcl.eblib.com/patron/FullRecord.aspx?p=4107798
Mazzucato, M. (2018). *Mission-oriented research & innovation in the European Union: A problem-solving approach to fuel innovation-led growth*. Publications Office of the European Union.
Wilsdon, J., & Rijcke, S. (2019). Europe the rule-maker. *Nature, 569*, 479–481. https://doi.org/10.1038/d41586-019-01568-x

Supplements

Real Knowledge please!
Dies Natalis, Dinner Speech, Utrecht University, 26 March, 2012.

Most progress in studying infectious disease (flu and AIDS) but also non-communicable diseases (AIZ, CVD, Alzheimer) is made when in multidisciplinary approaches the skills (techniques, concepts, logic) of distinct disciplines are integrated. As a result of this approach in Life Sciences, for instance, diseases that had nothing or very little in common 30 (15??) years ago are now believed to have common underlying pathogenic causes resulting from inflammatory processes.

It has become very clear that problems of the real world cannot be solved by research from within a single Life Science discipline (epidemiology, genetics, imaging neurology). Even more, to be successful and to significantly move the frontiers of science, high level convergence of medical engineering, chemistry, mathematics and life sciences has to be facilitated and achieved (Sharp). This means that science that aims to address real world - and hence complex- problems requires large scale multidisciplinary collaboration and large-scale investments. Principal investigators with their labs join large-team efforts, in concerted actions to approach pathogenesis, diagnosis and therapy of complex diseases. Since the early eighties, Life Science has been in transition to become Big Science, like Particle Physics has been since World war II. This is already going on for some time, but be aware, we have seen nothing yet....!

In addition to the shift to multidisciplinary research and convergence of engineering, physics and life science we have since the nineties seen the shift to what the commentators of science have designated **Mode-2 Science** (Novotny et al) or for which John Ziman coined the term **Post-Academic research.** This is the shift from curiosity-driven science to demand–driven research which starts when priorities have to be decided and the research agenda is set. It has been convincingly argued that science, not even pure science ever was value-free, but since science is an integrated part of society all its disguises predictably have major impact on our life and the life of billions around the globe, and of course the life of the generations to come.

F. Miedema, *Open Science: the Very Idea*,
https://doi.org/10.1007/978-94-024-2115-6

Our choice for the research agenda, given limited resources of talent, money, time and facilities, ideally has to reflect the needs of society. This is in principle a good thing since it may help to optimally engage science in addressing the most significant problems that we are facing. In that way science will produce societal, robust (Novotny et al) or as Philip Kitcher calls it **significant knowledge**. In this process of setting the research agenda's, which is of course imperfect even in our western democracies, the needs of the poor, the less powerful, children and future generations have to be actively tutored in order to balance their interests to the interests of the affluent parts of the world, of private parties with commercial interests and of lobby groups that are well-sponsored by special interest groups (Kitcher).

We have to realize that this may be hard for the 'ordinary' scientist who may feel threatened because she feels her scientific and professional autonomy affected. It may even be resisted by some from within their disciplinary turf, and this is not only true for medical disciplines, because they have to leave their comfort zone to work with these other MD's, engineers, preclinical scientists but also representatives from outside science to be eligible for the next round of funding. Indeed, to be able to do research **successfully and happily** in this new area of Life Science, social and management skills are required different from what was required when most of us started our PhD work in the late seventies or early eighties of the last century. It requires teamwork, deliberation with patients and patient-interest groups and increasingly private parties to decide which clinical needs have priority and which can be solved first and quickly or for which more basic work should be done. This truly is a co-creation process in which the client/patient, or as some like to call it 'stakeholder', and the developer/scientist work together from beginning to end. The ideal end product of course is implementation and evaluation of innovation in care or preventive medicine accompanied by scientific publications that duly report the findings to the community.

Educating Scientific Literacy

This needs explicit and implicit (tacit) training in the classroom and in labs and clinics. In order to get the required level of scientific literacy, we must expose our students to as much a possible multidisciplinarity in teaching, clinical care and research during their training in Medical School and the Life Sciences. The ethical problems of scientific choice, the almost mythical idea of value–free science, the involvement of values in all phases of the practice of science, the historical strong emphasis in the universities on basic and preclinical science at the cost of translational and applied science, the shift from individual to predominantly team work, the collaboration with societal stakeholders, it has to be put in the philosophical/ sociological context for our students because these issues determine the world in which they will be doing their science. We have to show them our excitement about the power and potential of modern science and help them to get a realistic picture of the practice of science, which as we all know is in all aspects human, muddy, political and imperfect but despite that produces fantastic reliable knowledge that can address our human needs (Kitcher; Miedema).

The Board Room

For us, the leaders and administrators in science, university, medical schools, etc this coming of age of modern- twenty- first- century science, for which I coined the name **Science 3.0**, has huge implications. They relate to the use of the so-called **credit cycle**, the economic reward system of science. In that system scientists produce data, that are written up and published and based on the quality of the journals where these publications appear, credit points are distributed (Stephan). Given the developments I just sketched, how should we recognize the contributions of individual scientists in such large scale multi-authored multidisciplinary 'productions'. A recent example from my former department: it took 3 PhD students, many lab techs, supervised by more than 3 PI's from different faculties and departments years of work to get a very nice ground- breaking paper in *Immunity,* a leading journal in our field.

How is credit to be dispersed? We have excellent researchers who produce not as much highly cited papers but design, develop and implement novel radiotherapeutic procedures to treat cancers. Techniques that are groundbreaking and significant because they address serious patient needs. Do we in the current system give proper credit for that type of work? It is a special interest for this audience to point to a recent thoughtful paper in Nature Immunology (!) by two leading American immunologists, Ron Germain and Pamela Schwartzenberg. They brought this problem 'of the changing sociology of academic translational research' up and urged us 'to define viable career paths' and 'special tenure track considerations for authors on such team-science publications'. Away with bean counting and back to content-based and science-informed professional management to steer our science? Although this may be much tougher than bean counting, I whole heartedly say yes! We must realize that it is the 'credit cycle' we use to manage research institutions, to give out or deny grants and accept papers, to decide on promotions and tenure and for hiring and firing Let's be honest here, these modern developments in the practice of science are a problem for the ' bean counters' that we have become, that the government has made us become, that -let's face it- we find sometimes handy and convenient.

It has been suggested that this reward system may be one of the causes for loss of actual quality of published work even in high impact journals (Begley and Ellis) which causes serious problems and loss of time and money for those who based on trust followed up on that work. The current system as it is may not be sustainable because it is focused on short term results which does not encourage risk taking and major breakthroughs but scientific projects with quick but incremental results. This works against the young scientists which is aggravated in particular by the increasingly unrealistic demands from reviewers and journal editors in order to increase quality and impact of their journals which does allow for a realistic development of career paths of young scientists (Ploegh, Stephan).

Here, I took the opportunity to paint the bigger picture of how science is rapidly developing, literally while we are watching. I pointed out what is expected from us in order to properly steer our institutions and to help facilitate career- development for our people, scientists and others. We have to come up with measures that regulate the 'free-market' in the system of science, modulating drivers and incentives of the various actors to ideally redirect it at the societal aims, avoiding the problems of the current system and make it sustainable. That could mean that we have to go, not for numbers of papers, impact factors, Hirsch factors, citation frequencies, the numbers of grants won in competition, but measures for societal (social and economic) significance.

Our academic community must realize that in this modern world of science, the public and government in return on investment want to see us producing Real Knowledge that addresses real felt societal needs and problems. I am convinced that can be done, it will not be easy, but it can be done if we work together.

Literature Cited

C.G. Begley en L.M. Ellis. Raise standards for preclinical cancer research, Nature 483 (2012) 531

Editorial. Creative tensions, Nature 484 (2012) 5

R.N. Germain en P.L.Schwartzberg. The human condition- an immunological perspective, Nature Immunology 12 (5) (2011) 369

S.Head. The Grim Threat to British Universities. New York Review of Books Januari, 2011

P. Kitcher, Science Truth and Democracy, Oxford University Press, Oxford, New York, 2001

F. Miedema. Science 3.0. Real Science Real Knowledge, Amsterdam University Press, 2012.

H.Nowotny, P.Scott en M.Gibbons, Re-thinking Science, Polity Press, 2001

H.L.Ploegh. End the wasteful tyranny of reviewer experiments, Nature **472** (2011) 391

P. Stephan. Perverse Incentives, Nature 484 (2012) 29

P.Sharp et al. The third revolution: The convergences of the Life Sciences, Physical Sciences and Engineering. MIT Januari 2011. (The Third Revolution: The Convergence of the Life Sciences, Physical Sciences and Engineering)

J. M. Ziman, *Real Science, What it is and what it means,* Cambridge University Press, Cambridge, 2000

My contribution to the First Science in Transition Workshop, with minor edits anno 2020 was:

The Enchanted View

Analyses of the historical, philosophical and sociological origins of the various stereotypes of science and scientists may help us understand where the enchanted view of science comes from. Here we always refer to positivistic and sometimes Popperian philosophy of science which is understood to implicate that science is about facts directly derived from experiments that can thus be rather objectively and directly verified. Based on this unique 'scientific method' science yields objective knowledge. The other important source is the Mertonian sociology: in order to see to it that scientists resist temptations, science is organized around the well-known Mertonian principles. Peer pressure is organized in a sociological system. Although

Merton's sociology was designed and required because scientists are human, para-doxically the Mertonian world view is romantic and idealized and free from (con-flict of) interests and politics.

It has been pointed out that Kuhn in *The Structure of Scientific Revolutions (SSR)* describes science as a system separated from the rest of society. However, he allows for psychological and sociological influences in debates about content, paradigm change is brought about by new data, but not without fierce negotiation between humans which involves all the game playing seen elsewhere in society. Economic interests coming from stakeholders outside are not explicitly referred to. In SSR professional conflicts of interest are obvious and are treated as normal and healthy in scientific debates, probably because Kuhn assumed that it was all taking place in a perfect Mertonian system were the players voluntarily adhered to the well-known values.

External values not directly relevant to the content or to the practice of science do not play a role, are not believed relevant and are not allowed. Issues of problem choice are thus treated as totally internal affairs to science and scientists. Problem choice therefore is not considered an issue apart from ideas that at every stage of a paradigm or field automatically a limited set of problems comes forward that the field agrees to be the top issues to be studied. Merton obviously, but also Kuhn, allows for elitism and stratification. Not all debaters are equal, but this is within the accepted way the game is played. The Matthew effect is pointed out as a sociologi-cal given, and not felt to be very problematic. This picture of science shows that science is intrinsically conservative, will resist change and innovation despite the Mertonian value system. Many have criticized Kuhn suggesting that he has written a normative and not a descriptive account of science because he seems to like the initial resistance to rejection of old theories and programs to allow for stable devel-opment and evolution of them avoiding loss of potential.

The enchanted view of science that results from these very influential descrip-tions or (to put it correctly) normative ideas about scientific activity is still largely the default mode when science and scientist are discussed in a public or more for-mal debate. The gossip and stories about clashes and fights between prominent sci-entists are of course all over the newspapers, TV and internet, and are enjoyed much, but the formal and official response is that those all are harmless because it is happening in 'Mertonian space'. As a consequence, there are a few issues that are increasingly felt to become problematic:

1. The idea that there is somehow always a high degree of consensus in science, or that it is possible at every given moment for any issue to generate that consensus for practical use by policy and public.
2. It is generally believed that problem choice is guided by 'an invisible hand ' determined by internal developments that we agree on based on the prevailing paradigms in a given field of research.

Ideological Use of the Enchanted View

It is relevant to analyse the use of this dominant stereotype in the different contexts, and specially for what purpose it is used and sustained. Of course, many especially those who are ignorant of the practice of modern science, naively teach and disseminate the classical view. This holds for high school teachers and even many who teach undergraduates. Those who have been exposed to the practice at the frontiers of science, often do not want to be or do not see why they should be, the cause of disillusion (disenchantment) that scare potential students away. In other more formal and public cases the enchanted view it seems is used with a political aim. This enhanced view is propagated and used to formally react to questions from outside science to defend science as unique and the only system we have to make reliable knowledge. The unique virtues of the system and its players, protection from non-scientific influences and interests and of course the scientific method are the defence lines. Fraud and bad science are in this reaction, universally treated as exceptions in an otherwise perfect system of self-cleansing peer review and post-publication criticism. All representative appearances of Robbert Dijkgraaf, the former president of the Royal Society, are typically in that style. Interestingly and in fact worrisome, he was never critically approached by a critical interviewer on this point. He was to all of us the ideal son-in-law and was given special programs on public TV. The Royal Society was most happy with four of these years of enchanted science. It was for instance well covered in the news that Dijkgraaf went to Princeton and Hans Clevers took over. It is believed that by sticking to this position and telling this version of science, public trust and trust at the part of key persons in administration and their representatives is maintained. It is felt that if we would tell the truth about how science works, how we know what we know, why we believe what we believe, how and to what extent interests shape our knowledge and scientific opinions, the public will lose faith and science will fall.

There thus seems to be an *omerta* (or, conspiracy of silence) regarding the practice of science. This has been shown early on by Gunther Stent in his wonderful analyses of the reviews of *The Double Helix* by Jim Watson. Gunther Stent wrote 1968 a *'Review of the Reviews' (Quarterly Review of Biology and Stent book)*. Stent documents why Watson was either cheered or reviled by respectable colleagues: he had as one of the top scientists broken 'the omerta'. The issue was not whether Watson had been unfair and critical about his colleagues and himself (!). The reviewers were embarrassed by the honesty and shamelessness by which he informs the reader how 'unscientifically' they behaved and how by all kind of sneaky tricks important parts of the critical data were gathered by Watson and Crick. *Wilkins wrote Watson that he also was 'tired of the polite covering up and misleading inadequate pictures of how scientific research is done',* but Watson had gone too far, showing the less mythical backstage practices of science. It is no secret that the first version of the book, which the was still entitled 'Honest Jim', had been seen by some of the reviewers and one of them thought the book had been *'bowdlerized here and there'* in the printed version.

Breaking the Omerta?

When we break the omerta, will we gain more in the long run than we lose short term? The use of these stereotypes may reassure some, but at the same time it confuses the public about what science has to offer to solve specific societal issues. For instance, it does not help to explain the debates and pluralism about climate change, the use of Flu and HPV vaccines, the battle against cancer, the coming epidemics of CVD and dementia, the cause and solution of the economic crisis, the approaches to multicultural societies, etc. It does not help to understand the interaction between scientific advisers and policy makers and governments. Sometimes scientists seem to agree, sometimes not, but it is unclear to the public why that is. It is difficult to explain to outsiders from within the classical view what type of pluralism can occur and for what reasons. The difference between professional and economic conflict of interest and how this affects integrity and trust. Will we be better able to explain bad science, bad pharma and fraud and that it will happen more and more by honestly explaining the system and its problems?

Questions to Be Discussed

How to avoid these stereotypes? How to exchange them by - in our view - a more realistic view of science and scientists? (if one can at least to some extent agree on that?) Can this be done without a further loss of trust in science and scientists? How to avoid a general distrust and loss of faith - and even nihilism - regarding modern science when we explain truthfully how science really works? How is knowledge constructed and produced in physics, experimental psychology, economics, life science, geology? How do we account for and explain the influence of internal and external forces and interests on science and still say that we believe that 'science can be trusted'? Or don't we? How do we think the well-known 'politics of science' can be explained without disturbing the trust and faith in science?

Science in Transition
September 25, 2013
The Position Paper of Science in Transition.
What should the science community do?

Discussants:
Prof dr Hans Clevers (president KNAW/ UMC Utrecht, University of Utrecht)
Prof dr Frank Miedema (Dean UMC Utrecht)
Prof dr Wijnand Mijnhardt (Director Descartes Centre, UU)
Dr Huub Dijstelbloem (WRR/University of Amsterdam)
Dr Karl Dittrich (VSNU)
Prof Dr Bert van der Zwaan (rector UU)
Prof Dr Andre Knottnerus (Director of the Netherlands Scientific Council for Government Policy, WRR)

Science in Transition
September 16, 2013
Conclusions & Recommendations of the Position Paper

Images of Science: We cherish an image in which scientists, through curiosity, provide undeniable knowledge. Knowledge society is in touch with. This image is however be incomplete. But besides incomplete it is even harmful. Any deviation of that image affects the trust of citizens in science and ultimately threatens the enterprise of science. While science is one of the driving forces of modern society, how does the image differ from reality? To begin with, science does not provide absolute certainty and consensus. Scientists can have different opinions. At the forefront of science rages a continuous conflict of professional interest, in which new knowledge is filtered in pruning debates. But where obsolete knowledge sometimes survives too long. In addition, scientists for producing interesting results are rewarded with reputation, promotion and sometimes personal economic gains. Scientists are ordinary people, with personal and social preferences, problems and needs. They sometimes do 'trimming and cooking of measurement data, get payments by industry, or just are not so good with methods in their field.

Recommendation: We need to inform the public about the uncertainty of scientific results, the way in which results are achieved and the everyday motives of scientists. This prevents theatrical public incomprehension of discussing scientists, about knowledge that does not prove to be true and misleading scientists.

Quality: Does the taxpayer get value for money? It is a valid question at a time when science is funded largely by public funds. The answer is unfortunately that many scientific results are more important for the scientist than for society. It is the result of the misuse of quantitative bibliometrics in the assessment of science. Scientists are judged by the number of publications in magazines with high impact factors. This makes publishing papers the highest goal of scientists. Whether they answer socially relevant questions is secondary. It also means that risky long-term research is hardly financed. In particular, life sciences have turned into a PhD factory. PhD students and postdocs do the bulk of the work, but without a lot of career prospects. But they will not hear from their mentors because they do not want to discourage their cheap labour forces.

Recommendation: Formulate new criteria for assessing scientists and scientific results, and emphasize the social value of the research emphatically.

Recommendation: Involve social stakeholders in the distribution of research money and in setting priorities in the research.

Trust: The public has high rust in the institute of science, more than in politics, journalism or business. But the time we trust a scientific "expert" is long gone. Their opinions are often contradictory. It can be traced to information crime and changed authority relations. The huge global knowledge production leads to hyper-specialization, resulting in a loss of overview. At the same time, the Internet makes it easier to interpret the information beyond the traditional frameworks. A range of public opinions is the result.

Recommendation: Be open and frank and show the public how science and research really is being done. At the front of research, both the natural and the social sciences, seriously different insights do exist, sometimes even for prolonged periods of time. Science has a lot to offer. Policy problems and their debates need science advice, although they are not decided but are informed by scientific knowledge as in the end not the scientists but the politicians decide. Scientific discussions very often also have moral, political and cultural aspects which relate to ideas and opinions about the world and how to life our lives.

Reliability & Corruption: The current organization of the scientific system puts pressure on the integrity of individual researchers. Emphasis on the number of published articles, plus the personal career motivations of scientists, means that the quality of research is under pressure. Many moderate, uninteresting, sometimes bad, and once-even fraudulent publications are produced. Although this is bad for science and society, those researchers need these papers to survive. In addition, more and more research takes place in cooperation with private parties. Such collaborations often yield useful results, but they also may create institutional and personal economic conflicts of interest. That is inevitable and also not fundamentally wrong, but it requires great vigilance

Recommendation: The interactions and financial dependence on third parties is associated with risks that can only be minimized by strict agreements in advance and strict supervision.Let public view how scientific decisions about social issues are taken. Make clear that the interests that play a role are not per se harmful, negotiation is part of the process. When speaking to researchers at congresses, in the public and in the media, one has to think that they also have personal motives and analyze and value their arguments based on it.

Communication: Communication departments of universities have been established in order to show the public how tax money is being used to do research that contributes to society. By default, the communications by these professionals aiming at the lay public are written in the logic of the 'ideal' university spreading the myth of certainty and absolute knowledge. Science journalism, pressed for time and limited by their own perverse incentives convert news from universities and the scientific journals into pretty short stories and footage. No room for nuances, uncertainty and depth.

Recommendation: Journalists should trained to expose and explain the mechanisms for science to the public which will increase mutual understanding. We have in mind investigative journalism such as how Joris Luyendijk researched the City of London about the origins of the financial crisis.

Democracy and Policy: Science has become an institutionalized capital-intensive social activity and must be treated as such. In a democratic society the public has the right to decide on the science agenda. For example, do we invest in the Higgs particle or in a malaria vaccine? Science is indispensable in political judgment and decision-making, and for informed social debate. However, her role is under pressure. Politicians and policy makers go selectively with findings from research. Job Search does not get too full since. Continuously there are experts who dispute the judgment of others. Social organizations come with counter-expertise. Science that wants to give advice is often politicized.

Recommendation: Both in basic and applied research, society must help to identify research priorities. Science can not scientifically determine its own course. This requires broad debates and considerations. The agenda of science is a matter of society. Contradictory insights about scientific research that want to advise policy and politics should not be heard in the background but on an audience stage. More often, researchers come to the House of Representatives, be less afraid of conflicting advice, try to find out where the differences are, cut any problems in parts if there is no overall solution. Allow experiments, learn from wrong paths.

A Crisis of the Entire University: Problems of production-driven scientific research not only play in life sciences and science, but also in the humanities and social sciences. The humanities discipline has set aside its task of educating teachers and is now focusing mainly on research. But the direct social justification for this is unclear and it produces many results that nobody's waiting for. Social sciences also play the main role in the international debate and are getting less attention from social issues in their own country. University width means that the ideal of higher education for many has gone out on a fiasco. There are good reasons to doubt the level of today's graduates. The quality of secondary education leaves much to be desired and many graduates have difficulty finding a job on a level. The perverse funding stimuli already explain a lot. If society increases the number of graduates, it should not be surprising that the quality per unit of product decreases. Moreover, the excessive growth of the number of students has undermined the university system too much.

Recommendation: Reinstate the university studies humanities and social sciences, focus less on research and more on education. Teaching as a professional profile for academics must be honored, with matching rewards.

Science in Transition Conference: November 7 and 8, 2013, KNAW Amsterdam

Over the next few years, science will have to make a number of important transitions. There is deeply-felt uncertainty and discontent on a number of aspects of the scientific system: the tools measuring scientific output, the publish-or-perish culture, the level of academic teaching, the scarcity of career opportunities for young scholars, the impact of science on policy, and the relationship between science, society and industry.

The checks and balances of our scientific system are in need of revision. To accomplish this, science should be evaluated on the basis of its added value to society. The public should be given a better insight in the process of knowledge production: what parties play a role and what issues are at stake? Stakeholders from society should become more involved in this process, and have a bigger say in the allocation of research funding. This is the view of the Science in Transition initiators Huub Dijstelbloem (WRR/UvA), Frank Huisman (UU/UM), Frank Miedema (UMC Utrecht), Jerry Ravetz (Oxford) and Wijnand Mijnhardt (Descartes Centre, UU).

Location: Tinbergenzaal, KNAW Trippenhuis, Kloveniersburgwal 29, Amsterdam

Suggested reading material: the SiT-position paper (in Dutch, or in English), and the Agenda for Change.

Program Thursday, 7 November

9:30–10:00 Registration, coffee and tea

10:00–10:30 Welcome, and historical perspective by Jerome Ravetz (Associate Fellow Saïd Business School, University of Oxford)

10:30–12:30 Morning session: Quality and corruption

Chair: Sally Wyatt (Professor of Digital Cultures in Development, Department Technology and Society Studies, Maastricht University)

Key note: Jan Vandenbroucke (Professor of Clinical Epidemiology, Department of Clinical Epidemiology, Leiden University Hospital)*

Commentators:

Henk van Houten (General Manager Philips Research)

Prof. Carl Moons, (professor in clinical epidemiology (UMC Utrecht)

Frank Miedema (Dean and vice-chairman of the Board, Professor of Immunology, UMC Utrecht)

12:30–14:00 Lunch, on own account

14:00–16:00 Afternoon session: Image and trust

Chair: Rob Hagendijk (Associate Professor Department Political Science, University of Amsterdam)

Key note: Sheila Jasanoff (Pforzheimer Professor of Science and Technology Studies, Harvard Kennedy School)

Commentators:

Ruud Abma (Assistant Professor Social Sciences Utrecht University / Descartes Centre)

Hans Altevogt (Greenpeace)

Jeroen Geurts (Chairman Young Academy KNAW, Professor Translational Neuroscience VU Medical Center)

Rudolf van Olden (Director Medical & Regulatory Glaxo Smith Kline Netherlands)

Frank Huisman (Professor in the History of Medicine, Julius Center UMC Utrecht / Descartes Centre)

16:00–16:30 Discussion

16:30 Drinks

*Drummond Rennie (Editor Journal of the American Medical Association, adjunct Professor of Medicine in the Institute for Health Policy Studies, University of California, San Francisco was scheduled, but unfortunately had to decline for health reasons the day before the meeting.

Science in Transition Conference, Day 2. 8 November 2013

Program Friday, 8 November

9:30–10:00 Registration, coffee and tea

10:00–10:15 Welcome

10:15–10:25 Column Maartje ter Horst (Student Universiteit Utrecht): *De harde waarheid: er zijn veel en veel te veel studenten*

10:25–12:30 Morning session: Communication and Democracy

Chair: Peter Vermij (Director Bird's Eye Communications)

Key note: Mark Brown (Professor in the Department of Government at California State University, Sacramento)

Commentators:

Peter Blom (CEO Triodos Bank)

Jasper van Dijk (Member of Parliament Socialist Party)

Rinie van Est (Coordinator Technology Assessment division Rathenau Institute)

Arthur Petersen (Chief Scientist at the PBL Netherlands Environmental Assessment Agency (Planbureau voor de Leefomgeving, PBL) / Professor of Science and Environmental Public Policy in the IVM Institute for Environmental Studies at VU University Amsterdam)

Huub Dijstelbloem (Senior staff member Scientific Council for Government Policy (WRR) / Department Philosophical Tradition in Context, University of Amsterdam)

12:30–14:00 Lunch on own account

14:00–14:10 Column Hendrik Spiering (Chef Wetenschap NRC): *Nieuwe tijden, nieuwe wetenschap*

14:10–16:00 Afternoon session: General debate and conclusions

Chair: Mirko Noordegraaf (Professor of Public Management, Utrecht School of Governance, Utrecht University)

Key note: Hans Clevers (President of the Royal Netherlands Academy of Arts and Sciences (KNAW))

Commentators:

Jos Engelen (Chairman Netherlands Organisation for Scientific Research (NWO))

André Knottnerus (Chairman Scientific Council for Government Policy (WRR))

Lodi Nauta (Dean Faculty of Philosophy, Professor in History of Philosophy, University of Groningen)

Wijnand Mijnhardt (Director Descartes Centre for the History and Philosophy of the Sciences and the Humanities / Professor Comparative History of the Sciences and the Humanities, Utrecht University)

16:00- 16:30 Discussion

16:30 Drinks

A Toolbox for Science in Transition

17 September 2013, KNAW Symposium about Trust in Science
Reflections

1. **Science in Transition: a science of science project.**

   The first tool we need is comprehensive scientific research on the current system of science. This involves multidisciplinary theoretical, but also empirical work mainly on the sociology and economy of science, but also historical analyses of the changes that science has gone through, since World War II. This project has started @SciTransit, website www.

   Tools to start with now:

2. Impact

The evaluation of research impact needs to change. We have to move on from the now used intransience, self -referential metrics to integral evaluation. Measures both for in-science use and for societal impact, value attributed by potential users in society must be developed. Although many colleagues are still very cautious, this process has started already. Much ideas and material are being produced in several countries, including but especially UK and in NL. KNAW, VSNU and NFU are involved in what in the NFU is known as the 'Impact' project. At UMC Utrecht we are moving to hybrid fora, inviting representatives from outside science to partici-pate in research evaluation and to more integral metrics in science excellence evaluation.

Internationally, the simple use of impact factors is on the way out, see the San Francisco Declaration on Research Assessment. This and the evolution of altmetrics will challenge the standard practice of commercial journals and they know it and are in anticipation taking all kinds of action.

3. Incentives and rewards

This comprehensive measurement of research impact thus has forward looking effects. Based on these '3.0 evaluations', we must manage and facilitate science. Research management will have to take into account the various types of impact and provide career opportunities for the different types of researchers. This is going on here and there already but must become the dominant practice in public research management.

Grants from the respective institutions, NWO, ZONMW and charities should explicitly be awarded based on the same principles. The classical metrics using IF, citations and H factors should be complemented by an accepted list of measures of societal impact. This must result in a mix of basic and targeted research, where also basic science is judged in a wider context, since also pure basic science is a 'politi-cal' choice

4. Dealing with Risk Avoidance

NWO, ERC and other high-profile personal grants tend to select classical researchers, based on the usual metrics, that produce knowledge for the internal sci-ence market mainly. Because of risk reduction, notwithstanding what the members of these panels say, and tend to believe, too much emphasis is on high impact pub-lications. This results in too little diversification of leading investigators in the institutes.

In addition, there is a skewed emphasis and overrated valuation of these laureates over many other non-laureates who are very good researchers as well, but just were not selected because of lack of funds at NWO or ERC. Many of these should be brought in university talent programs which may and should select for quite differ-ent capacities that are not detectable at a snapshot 10 minutes interview and from classical metrics. Universities are able to go for less risk avoidance and long-term innovation and may and should invest in that. Real world problems are complex and require intense collaboration by several disciplines. Obviously, we have to work from and with realistic expectations regarding research output. If we start from excessive demands, researchers will be tempted to play the system in order to survive to the next level.

5. **PhD Talent management and education**

As a recent Rathenau report bravely pointed out, granting organizations should give grants, but the university must take up the talent and career management of its staff, because getting prestigious grants clearly is only for the top 3%, but for the top 30% getting funded is a random and subjective process indeed. Talent management in general is a problem given the huge PhD factories we have established and the poor prospects these PhDs and postdocs have to make a career in academia. Indeed as Solla Price 50 years ago predicted, science breeds scientists faster than economy wants to pay for. See Stephan's recent book 'How economics shapes Science'. We have to rethink and redesign these streams to better coach the careers of our students for careers in and outside science. This is a major challenge for universities. How much PhD should we produce? Do MD's in those high numbers need a PhD at all. How are we to promote transdisciplinary work via the talent programs? We have to promote scientific literacy among our Master and PhD students by courses like 'That Thing Called Science' that we started in Utrecht in 2009. This education of scientific literacy has recently started in many of our universities.

6. **Targeting societal problems**

As the KNAW advice Vertrouwen/Trust correctly points out, trust in science and scientists is not only about how trustworthy and careful we do and report our science, it is as much about the question whether we do the rights things. For charities and other targeted science funding this is not problematic. These funders are already changing practice in that they steer for large problem-oriented multidisciplinary programs involving teams from several institutes. This in most cases is a mix of lab and patient-oriented translational work.

For universities that fortunately still have their own intramural funding, this calls for change. They have to relate more with stakeholders in society to (re)orient their research, both basic and targeted in a mode-2 fashion. Not only in Utrecht this has already started to varying degrees, in anticipation of EU HORIZON 2020 which targets the so-called Societal Grand Challenges.

Also, NWO has been advised by a recent committee to take this programmatic/thematic orientation. NWO because of its inward looking, rather positivistic & reductionist orientation is not trusted by societal partners for starting programs on real world problems. This of course is played out not even subtle by parties from the private sector who want to try to secure a bigger piece of the pie. It is to me a question still, whether and how NWO could do programming of science.

7. **Communication.**

Finally, how to tell and sell the story of modern science to improve our relations with and trust by the public. My proposal is to be honest and open, also to our students and young researchers about how science in practice really works. Let's do away with our stories about the myth of science as a perfect method performed by individuals with high moral values without any bias or interests. Let's explain how science does make objective facts but that a lot of uncertainty remains. Let's tell about science as a job, as a career and how the economy of science shapes science and the content of our science. Be honest, that in particular regarding complex real-world problems a lot of uncertainty remains and that scientists have personal beliefs that shape their scientific ideas. I recommend Paul Wouters' paper in Dijstelbloem and Hagendijk's book.

8. **Problem choice**

Finally, problem of agenda setting in a democratic society presents itself. How are we organizing this process designated by Kitcher the 'ideal deliberation'? In our country, as probably elsewhere, this is an imperfect and haphazard process that is not really transparent and not really democratic. How did we arrive at the TOP SECTORS? How ideal were these deliberations? Is there in the future a role for the KNAW? The KNAW has produced in 2011 a little book with 49 big questions coming up from the scientific fields, as sort of a national research agenda. This was, however, not in deliberation with representatives from society. Much change is needed but this is very complex indeed.

Supplement 2

The Metric Tide, 2015

Recommendations
Supporting the effective leadership, governance and management of research cultures

1 **The research community should develop a more sophisticated and nuanced approach to the contribution and limitations of quantitative indicators**. Greater care with language and terminology is needed. The term 'metrics' is often unhelpful; the preferred term 'indicators' reflects a recognition that data may lack specific relevance, even if they are useful overall. (HEIs, funders, managers, researchers)

2 **At an institutional level, HEI leaders should develop a clear statement of principles on their approach to research management and assessment, including the role of quantitative indicators.** On the basis of these principles, they should carefully select quantitative indicators that are appropriate to their institutional aims and context. Where institutions are making use of league tables and ranking measures, they should explain why they are using these as a

means to achieve particular ends. Where possible, alternative indicators that support equality and diversity should be identified and included. Clear communication of the rationale for selecting particular indicators, and how they will be used as a management tool, is paramount. As part of this process, HEIs should consider signing up to DORA, or drawing on its principles and tailoring them to their institutional contexts. (Heads of institutions, heads of research, HEI governors)

3 **Research managers and administrators should champion these principles and the use of responsible metrics within their institutions.** They should pay due attention to the equality and diversity implications of research assessment choices; engage with external experts such as those at the Equality Challenge Unit; help to facilitate a more open and transparent data infrastructure; advocate the use of unique identifiers such as ORCID iDs; work with funders and publishers on data interoperability; explore indicators for aspects of research that they wish to assess rather than using existing indicators because they are readily available; advise senior leaders on metrics that are meaningful for their institutional or departmental context; and exchange best practice through sector bodies such as ARMA. (Managers, research administrators, ARMA)

4 **HR managers and recruitment or promotion panels in HEIs should be explicit about the criteria used for academic appointment and promotion decisions.** These criteria should be founded in expert judgement and may reflect both the academic quality of outputs and wider contributions to policy, industry or society. Judgements may sometimes usefully be guided by metrics, if they are relevant to the criteria in question and used responsibly; article-level citation metrics, for instance, might be useful indicators of academic impact, as long as they are interpreted in the light of disciplinary norms and with due regard to their limitations. Journal-level metrics, such as the JIF, should not be used. (HR managers, recruitment and promotion panels, UUK)

5 **Individual researchers should be mindful of the limitations of particular indicators in the way they present their own CVs and evaluate the work of colleagues.** When standard indicators are inadequate, individual researchers should look for a range of data sources to document and support claims about the impact of their work. (All researchers)

6 **Like HEIs, research funders should develop their own context-specific principles for the use of quantitative indicators in research assessment and management** and ensure that these are well communicated, easy to locate and understand. They should pursue approaches to data collection that are transparent, accessible, and allow for greater interoperability across a diversity of platforms. (UK HE Funding Bodies, Research Councils, other research funders)

7 **Data providers, analysts and producers of university rankings and league tables should strive for greater transparency and interoperability between different measurement systems.** Some, such as the Times Higher Education (THE) university rankings, have taken commendable steps to be more open about their choice of indicators and the weightings given to these, but other

rankings remain 'black- boxed'. (Data providers, analysts and producers of university rankings and league tables)

8 **Publishers should reduce emphasis on journal impact factors as a promotional tool, and only use them in the context of a variety of journal-based metrics that provide a richer view of performance.** As suggested by DORA, this broader indicator set could include 5-year impact factor, EigenFactor, SCImago, editorial and publication times. Publishers, with the aid of Committee on Publication Ethics (COPE), should encourage responsible authorship practices and the provision of more detailed information about the specific contributions of each author. Publishers should also make available a range of article-level metrics to encourage a shift toward assessment based on the academic quality of an article rather than JIFs. (Publishers)

9 **There is a need for greater transparency and openness in research data infrastructure. A set of principles should be developed for technologies, practices and cultures that can support open, trustworthy research information management.** These principles should be adopted by funders, data providers, administrators and researchers as a foundation for further work. (UK HE Funding Bodies, RCUK, Jisc, data providers, managers, administrators)

10 **The UK research system should take full advantage of ORCID as its preferred system of unique identifiers.** ORCID iDs should be mandatory for all researchers in the next REF. Funders and HEIs should utilise ORCID for grant applications, management and reporting platforms, and the benefits of ORCID need to be better communicated to researchers. (HEIs, UK HE Funding Bodies, funders, managers, UUK, HESA)

11 **Identifiers are also needed for institutions, and the most likely candidate for a global solution is the ISNI, which already has good coverage of publishers, funders and research organisations.** The use of ISNIs should therefore be extended to cover all institutions referenced in future REF submissions, and used more widely in internal HEI and funder management processes. One component of the solution will be to map the various organisational identifier systems against ISNI to allow the various existing systems to interoperate. (UK HE Funding Bodies, HEIs, funders, publishers, UUK, HESA)

12 **Publishers should mandate ORCID iDs and ISNIs and funder grant references for article submission, and retain this metadata throughout the publication lifecycle.** This will facilitate exchange of information on research activity, and help deliver data and metrics at minimal burden to researchers and administrators. (Publishers and data providers)

13 **The use of digital object identifiers (DOIs) should be extended to cover all research outputs.** This should include all outputs submitted to a future REF for which DOIs are suitable, and DOIs should also be more widely adopted in internal HEI and research funder processes. DOIs already predominate in the journal publishing sphere – they should be extended to cover other outputs where no identifier system exists, such as book chapters and datasets. (UK HE Funding Bodies, HEIs, funders, UUK)

14 **Further investment in research information infrastructure is required.** Funders and Jisc should explore opportunities for additional strategic investments, particularly to improve the interoperability of research management systems. (HM Treasury, BIS, RCUK, UK HE Funding Bodies, Jisc, ARMA)

Recommendation
Research funders need to increase investment in the science of science policy. There is a need for greater research and innovation in this area, to develop and apply insights from computing, statistics, social science and economics to better understand the relationship between research, its qualities and wider impacts. (Research funders)

Supplement 3

The Leiden Manifesto

Ten principles[1]

1. **Quantitative evaluation should support qualitative, expert assessment.** Quantitative metrics can challenge bias tendencies in peer review and facilitate deliberation. This should strengthen peer review, because making judgements about colleagues is difficult without a range of relevant information. However, assessors must not be tempted to cede decision-making to the numbers. Indicators must not substitute for informed judgement. Everyone retains responsibility for their assessments.
2. **Measure performance against the research missions of the institution, group or researcher.** Programme goals should be stated at the start, and the indicators used to evaluate performance should relate clearly to those goals. The choice of indicators, and the ways in which they are used, should take into account the wider socio-economic and cultural contexts. Scientists have diverse research missions. Research that advances the frontiers of academic knowledge differs from research that is focused on delivering solutions to societal problems. Review may be based on merits relevant to policy, industry or the public rather than on academic ideas of excellence. No single evaluation model applies to all contexts.
3. **Protect excellence in locally relevant research.** In many parts of the world, research excellence is equated with English-language publication. Spanish law, for example, states the desirability of Spanish scholars publishing in high-impact journals. The impact factor is calculated for journals indexed in the US-based and still mostly English-language Web of Science. These biases are particularly problematic in the social sciences and humanities, in which research is more regionally and nationally engaged. Many other fields have a national or regional dimension — for instance, HIV epidemiology in sub-Saharan Africa.

---

[1] Source: Hicks, D., Wouters, P., Waltman, L., de Rijcke, S., & Rafols, I. (2015). Bibliometrics: The Leiden Manifesto for research metrics. *Nature, 520* (7548), 429–431. https://doi.org/10.1038/520429a

This pluralism and societal relevance tends to be suppressed to create papers of interest to the gatekeepers of high impact: English-language journals. The Spanish sociologists that are highly cited in the Web of Science have worked on abstract models or study US data. Lost is the specificity of sociologists in high-impact Spanish-language papers: topics such as local labour law, family health care for the elderly or immigrant employment. Metrics built on high-quality non-English literature would serve to identify and reward excellence in locally relevant research.

4. **Keep data collection and analytical processes open, transparent and simple.** The construction of the databases required for evaluation should follow clearly stated rules, set before the research has been completed. This was common practice among the academic and commercial groups that built bibliometric evaluation methodology over several decades. Those groups referenced protocols published in the peer-reviewed literature. This transparency enabled scrutiny. For example, in 2010, public debate on the technical properties of an important indicator used by one of our groups (the Centre for Science and Technology Studies at Leiden University in the Netherlands) led to a revision in the calculation of this indicator. Recent commercial entrants should be held to the same standards; no one should accept a black-box evaluation machine.
Simplicity is a virtue in an indicator because it enhances transparency. But simplistic metrics can distort the record (see principle 7). Evaluators must strive for balance — simple indicators true to the complexity of the research process.
"Simplicity is a virtue in an indicator because it enhances transparency."

5. **Allow those evaluated to verify data and analysis.** To ensure data quality, all researchers included in bibliometric studies should be able to check that their outputs have been correctly identified. Everyone directing and managing evaluation processes should assure data accuracy, through self-verification or third-party audit. Universities could implement this in their research information systems and it should be a guiding principle in the selection of providers of these systems. Accurate, high-quality data take time and money to collate and process. Budget for it.

6. **Account for variation by field in publication and citation practices.** Best practice is to select a suite of possible indicators and allow fields to choose among them. A few years ago, a European group of historians received a relatively low rating in a national peer-review assessment because they wrote books rather than articles in journals indexed by the Web of Science. The historians had the misfortune to be part of a psychology department. Historians and social scientists require books and national-language literature to be included in their publication counts; computer scientists require conference papers be counted.
Citation rates vary by field: top-ranked journals in mathematics have impact factors of around 3; top-ranked journals in cell biology have impact factors of about 30. Normalized indicators are required, and the most robust normalization method is based on percentiles: each paper is weighted on the basis of the percentile to which it belongs in the citation distribution of its field (the top 1%, 10% or 20%, for example). A single highly cited publication slightly improves

the position of a university in a ranking that is based on percentile indicators, but may propel the university from the middle to the top of a ranking built on citation averages.

7. **Base assessment of individual researchers on a qualitative judgement of their portfolio.** The older you are, the higher your h-index, even in the absence of new papers. The h-index varies by field: life scientists top out at 200; physicists at 100 and social scientists at 20–30 (ref. 8). It is database dependent: there are researchers in computer science who have an h-index of around 10 in the Web of Science but of 20–30 in Google Scholar. Reading and judging a researcher's work is much more appropriate than relying on one number. Even when comparing large numbers of researchers, an approach that considers more information about an individual's expertise, experience, activities and influence is best.

8. **Avoid misplaced concreteness and false precision.** Science and technology indicators are prone to conceptual ambiguity and uncertainty and require strong assumptions that are not universally accepted. The meaning of citation counts, for example, has long been debated. Thus, best practice uses multiple indicators to provide a more robust and pluralistic picture. If uncertainty and error can be quantified, for instance using error bars, this information should accompany published indicator values. If this is not possible, indicator producers should at least avoid false precision. For example, the journal impact factor is published to three decimal places to avoid ties. However, given the conceptual ambiguity and random variability of citation counts, it makes no sense to distinguish between journals on the basis of very small impact factor differences. Avoid false precision: only one decimal is warranted.

9. **Recognize the systemic effects of assessment and indicators.** Indicators change the system through the incentives they establish. These effects should be anticipated. This means that a suite of indicators is always preferable — a single one will invite gaming and goal displacement (in which the measurement becomes the goal). For example, in the 1990s, Australia funded university research using a formula based largely on the number of papers published by an institute. Universities could calculate the 'value' of a paper in a refereed journal; in 2000, it was Aus$800 (around US$480 in 2000) in research funding. Predictably, the number of papers published by Australian researchers went up, but they were in less-cited journals, suggesting that article quality fell.

10. **Scrutinize indicators regularly and update them.** Research missions and the goals of assessment shift and the research system itself co-evolves. Once-useful metrics become inadequate; new ones emerge. Indicator systems have to be reviewed and perhaps modified. Realizing the effects of its simplistic formula, Australia in 2010 introduced its more complex Excellence in Research for Australia initiative, which emphasizes quality.

Next steps

Abiding by these ten principles, research evaluation can play an important part in the development of science and its interactions with society. Research metrics can provide crucial information that would be difficult to gather or understand by means of individual expertise. But this quantitative information must not be allowed to morph from an instrument into the goal.

The best decisions are taken by combining robust statistics with sensitivity to the aim and nature of the research that is evaluated. Both quantitative and qualitative evidence are needed; each is objective in its own way. Decision-making about science must be based on high-quality processes that are informed by the highest quality data.

Supplement 4

Amsterdam Call for Action on Open Science

Amsterdam conference on 'Open Science – From Vision to Action' hosted by the Netherlands' EU presidency on 4 and 5 April 2016.

Open science Open science is about the way researchers work, collaborate, interact, share resources and disseminate results. A systemic change towards open science is driven by new technologies and data, the increasing demand in society to address the societal challenges of our times and the readiness of citizens to participate in research.

Increased openness and rapid, convenient and high-quality scientific communication - not just among researchers themselves but between researchers and society at large - will bring huge benefits for science itself, as well as for its connection with society.

Open science has impact and has the potential to increase the quality and benefits of science by making it faster, more responsive to societal challenges, more inclusive and more accessible to new users. An example of this potential is the response to the outbreak of viral diseases such as Ebola and Zika. Access to the most recent scientific knowledge for a broad group of potential contributors, including new or unknown users of knowledge, has brought solutions closer. Open science also increases business opportunities.

The speed at which innovative products and services are being developed is steadily increasing. Only companies (notably SMEs), entrepreneurs and innovative young people that have access to the latest scientific knowledge are able to apply this knowledge and to develop new market possibilities.

Citizen science brings research closer to society and society closer to research.

A Speedy Transition Is Needed

For Europe to remain at the forefront and to ensure sustainable growth in the future, open science holds many promises. Reality, however, has not caught up yet with the emerging possibilities. The majority of scientific publications, research data and other research outputs are not freely accessible or reusable for potential users. Assessment, reward and evaluation systems in science are still measuring the old way.

Although these issues are recognised and countless initiatives have been developed during recent years, policies are not aligned, and expertise can be shared more and better. There is a strong need for cooperation, common targets, real change, and stocktaking on a regular basis for a speedy transition towards open science.

The good news is that there is political and societal momentum. More and more researchers are supporting the transition and are moving towards open science in the way they work. Organisations from the scientific community are urging politicians to act. The European Commission and the Council of the European Union have expressed that they are prepared to take a leading role to facilitate and accelerate the transition towards open science.

From Vision to Action

This Call for Action is the main result of the Amsterdam conference on 'Open Science – From Vision to Action' hosted by the Netherlands' EU presidency on 4 and 5 April 2016. It is a living document reflecting the present state of open science evolution. Based on the input of all participating experts and stakeholders* as well as outcomes of preceding international meetings and reports, a multi-actor approach was formulated to reach two important pan-European goals for 2020:

Full Open Access for All Scientific Publications
This requires leadership and can be accelerated through new publishing models and compliance with standards set.

1. **A fundamentally new approach towards optimal reuse of research data**
   Data sharing and stewardship is the default approach for all publicly funded research. This requires definitions, standards and infrastructures.
   To reach these goals by 2020 we need flanking policy:

2. **New assessment, reward and evaluation systems**
   New systems that really deal with the core of knowledge creation and account for the impact of scientific research on science and society at large, including the economy, and incentivise citizen science.

3. **Alignment of policies and exchange of best practices**
   Practices, activities and policies should be aligned and best practices and information should be shared. It will increase clarity and comparability for all parties concerned and help to achieve joint and concerted actions. This should be accompanied by regular monitoring-based stocktaking.

Twelve Action Items with Concrete Actions to Be Taken

Twelve action items have been included in this Call for Action. They all contribute to the transition towards open science and have been grouped around five cross- cutting themes that follow the structure of the European Open Science Agenda as proposed by the European Commission. This may help for a quick-start of the Open Science Policy Platform that will be established in May 2016. Each action item contains concrete actions that can be taken immediately by the Member States, the European Commission and the stakeholders.

ACTION 1. Change assessment, evaluation and reward systems in science

The Problem

Open science presents the opportunity to radically change the way we evaluate, reward and incentivise science. Its goal is to accelerate scientific progress and enhance the impact of science for the benefit of society. By changing the way we share and evaluate science, we can provide credit for a wealth of research output and contributions that reflect the changing nature of science.

The assessment of research proposals, research performance and researchers serves different purposes, but often seems characterised by a heavy emphasis on publications, both in terms of the number of publications and the prestige of the journals in which the publications should appear (citation counts and impact factor). This emphasis does not correspond with our goals to achieve societal impact alongside scientific impact. The predominant focus on prestige fuels a race in which the participants compete on the number of publications in prestigious journals or monographs with leading publishers, at the expense of attention for high-risk research and a broad exchange of knowledge. Ultimately this inhibits the progress of science and innovation, and the optimal use of knowledge.

The Solution

- Ensure that national and European assessment and evaluation systems encourage open science practices and timely dissemination of all research outputs in all phases of the research life cycle.
- Create incentives for an open science environment for individual researchers as well as funding agencies and research institutes.
- Acknowledge the different purposes of evaluation and what 'right' criteria are. Amend national and European assessment and evaluation systems in such a way that the complementary impact of scientific work on science as well as society at large is taken into account.

Engage researchers and other key stakeholders, including communications platforms and publishers within the full spectrum of academic disciplines. Set up assessment criteria and practices, enabling researchers to exactly understand how they will be assessed and that open practices will be rewarded.

Concrete Actions

- **National authorities and the European Commission:** acknowledge that national initiatives are reaching their limits, and that this is an area for a harmonised EU approach.
- **National authorities, European Commission and research funders:** reform reward systems, develop assessment and evaluation criteria, or decide on the selection of existing ones (e.g. DORA for evaluations and the Leiden Manifesto for research metrics), and make sure that evaluation panels adopt these new criteria.
- **Research Performing Organisations, research funders and publishers:** further facilitate and explore the use of so-called alternative metrics where they appear adequate to improve the assessment of aspects such as the impact of research results on society at large. Experiment with new approaches for rewarding scientific work.
- **Research communities, research funders and publishers:** develop and adopt citation principles for publications, data and code, and other research outputs, which include persistent identifiers, to ensure appropriate rewards and acknowledgment of the authors.
- **Research communities and publishers:** facilitate and develop new forms of scientific communication and the use of alternative metrics.

Expected Positive Effects

- An end to the vicious circle that forces scientists to publish in ever more prestigious journals or monographs and reinforcement of the recognition for other forms of scientific communication;
- A wider dissemination of a wider range of scientific information that benefits not only science itself but society as a whole, including the business community;
- A better return for the parties that fund research.

https://www.government.nl/documents/reports/2016/04/04/amsterdam-call-for-action-on-open-science

Supplement 5

Rome Declaration on Responsible Research and Innovation in Europe

November 2014

Responsible Research and Innovation (RRI) is the on-going process of aligning research and innovation to the values, needs and expectations of society. Decisions in research and innovation must consider the principles on which the European Union is founded, i.e. the respect of human dignity, freedom, democracy, equality, the rule of law and the respect of human rights, including the rights of persons belonging to minorities.

RRI requires that all stakeholders including civil society are responsive to each other and take shared responsibility for the processes and outcomes of research and innovation. This means working together in: science education; the definition of research agendas; the conduct of research; the access to research results; and the application of new knowledge in society- in full respect of gender equality, the gender dimension in research and ethics considerations.

More than a decade of research and pilot activities on the interplay between science and society points to three main findings. First, we cannot achieve technology acceptance by way of good marketing. Second, diversity in research and innovation as well as the gender perspective is vital for enhancing creativity and improving scientific quality. And third, early and continuous engagement of all stakeholders is essential for sustainable, desirable and acceptable innovation. Hence, excellence today is about more than ground-breaking discoveries – it includes openness, responsibility and the co-production of knowledge.

The benefits of Responsible Research and Innovation go beyond alignment with society: it ensures that research and innovation deliver on the promise of smart, inclusive and sustainable solutions to our societal challenges; it engages new perspectives, new innovators and new talent from across our diverse European society, allowing to identify solutions which would otherwise go unnoticed; it builds trust between citizens, and public and private institutions in supporting research and innovation; and it reassures society about embracing innovative products and services; it assesses the risks and the way these risks should be managed.

European regions and countries are already engaged in this approach. Societal demands for ambitious environmental policies led to creative social and technological innovations such as fuel efficient vehicles, solar devices or mobility and recycling solutions based on sharing.

Therefore, we, the participants and organisers of the conference "Science, Innovation and Society: achieving Responsible Research and Innovation" held in Rome on 19–21 November 2014 under the auspices of the Italian Presidency, consider it as our collective duty to further promote Responsible Research and Innovation in an integrated way.

We call on European Institutions, EU Member States and their R&I Funding and Performing Organisations, business and civil society to make Responsible Research and Innovation a central objective across all relevant policies and activities, including in shaping the European Research Area and the Innovation Union.

The present declaration builds on the 2009 Lund Declaration, which called for an emphasis on societal challenges, and on the 2013 Vilnius Declaration, which underlined that a resilient partnership with all relevant actors is required if research is to serve society.

We believe the conditions are now right for responsible research and innovation to underpin European research and innovation endeavour and therefore call on all stakeholders to work together for inclusive and sustainable solutions to our societal challenges.

A description of the six dimensions of RRI can be found on http://ec.europa.eu/research/science-society/document_library/pdf_06/responsible-research-and-innovation-leaflet_en.pdf

Supplement 6

Summary Article MLE Open Science – Enabling Systemic Change Through Mutual Learning

Small fixes are not enough to reach Open Science's full potential. Systemic and comprehensive change in science governance and evaluation is needed across the EU and beyond, report experts in a recent Policy Support Facility mutual learning exercise. As a truly global movement, Open Science strives to improve accessibility to and reusability of research practices and outcomes. But the benefits of Open Science touch almost every aspect of society, including the economy, social innovation, and wider sustainable development goals.

> Open Science is more than Open Access and Open Data; it is a way of looking at the world, with the intent of building a better society. Bart Dumolyn, Policy Advisor on Open Science and Responsible Research and Innovation for the Flemish Government.

In its broadest definition, Open Science covers Open Access to publications, Open Research Data and Methods, Open Source Software, Open Educational Resources, Open Evaluation, and Citizen Science. But openness also means making the scientific process more inclusive and accessible to all relevant actors, within and beyond the scientific community.

With its many initiatives and programmes, Europe has long championed Open Science practices as a powerful means and excellent opportunity to renegotiate the social roles and responsibilities of publicly-funded research – and to rethink the science system as a whole.

The Horizon 2020 Policy Support Facility (PSF) gives Member States and Associated Countries the opportunity to request and take part in mutual learning exercises (MLE) addressing specific research and innovation policy challenges. The transition to Open Science represents such a policy challenge which is best tackled in close cooperation with all stakeholders and on an international scale.

Given that there is no common baseline for how to implement Open Science nationally, the MLE embraced a hands-on, 'learning by doing' approach supported by external expertise. Concrete examples, models, best practices and knowledge exchanges fostered broader understanding of the implications and benefits of Open Science strategies.

Problems and concerns were discussed in an 'open' and constructive fashion. The final PSF report. entitled 'Mutual Learning Exercise on Open Science: Altmetrics and Rewards', builds on this rich exchange of experiences, both positive and negative, and provides an overview of various approaches to Open Science implementation across Europe, which include different stakeholders and research communities.

MLE participants agreed that small fixes are not enough: implementing Open Science requires systemic and comprehensive change in science governance and evaluation. Crucial for a successful transition to Open Science will be strategic shifts in the incentives and reward systems.

Key Lessons on the Transition to Open Science

The scope of this first MLE on Open Science was narrowed down to address three topics, all of which are key elements of the European Open Science Agenda:

1. The potential of altmetrics – alternative (i.e. non-traditional) metrics that go beyond citations of articles – to foster Open Science
2. Incentives and rewards for researchers to engage in Open Science activities
3. Guidelines for developing and implementing national policies for Open Science

There can be no mission- oriented approach to research and innovation without Open Science. Michalis Tzatzanis, Austrian Research Promotion Agency (FFG).

Many MLE participants voiced concerns that altmetrics may encourage a business-as-usual scenario, with users focusing only on what is measurable and ending up with proxies far too simplistic for decision-making. Generally it was agreed that altmetrics have the potential to foster a major shift in the way research activities are evaluated and rewarded, providing they are open and reproducible in their method and data, as well as clearly indicate what qualities they measure.

So, what research qualities and societal benefits matter the most, how can they be tracked and measured, and for what reasons? Altmetrics can only help to break away from traditional indicators and publishing avenues, and establish themselves as responsible metrics if they cover diverse types of research practices and out-comes, according to the report, instead of "overly-simplified one-stop shops". Here, the MLE confirmed the concerns and recommendations put forward by a dedicated

Expert Group on Altmetrics and endorsed the coming activities of a European Forum for Next Generation Metrics.

MLE participants further called for clear goals and missions against which Open Science should be evaluated. Based on cross-national exchanges in the use of altmetrics in policy, the report called for more research on how they could be used not only to promote openness, but also as tools for more profound change – diversifying innovation landscapes and raising awareness of niche pockets of excellence. Altmetrics could also provide visible links between education and science and help to overcome the problem of research fragmentation across Europe and beyond.

> Participation in the MLE provided a great opportunity to get closer and deeper insight into the implementation of various practices of Open Science. The established contacts and information provided encouraged me to propose concrete measures to our leaders. Aušra Gribauskiene, Chief Officer of the Science Division of the Ministry of Education and Science of the Republic of Lithuania.

It is extremely difficult for researchers to adopt Open Science practices without a broad institutional shift in support and evaluation structures governing their work. Discussions during the MLE revealed that very few Open Science **incentives and rewards** are currently being implemented in participating countries. MLE participants underlined the necessity to develop incentives for different stakeholders: researchers, research organizations and funders, national governments and policymakers.

Since incentives for researchers need to include radical shifts in hiring and promotion procedures, a very good blueprint for future approaches is the Open Science Career Assessment Matrix (OSCAM). This scheme details the different ways that researchers' less visible work and other types of research outputs can be acknowledged or measured.

Given the highly international nature of research networks, international coordination is crucial to the effective implementation of comparable measures. Each country, research funder and research- performing organisation needs to review the extent to which specific incentives will work on the ground, and adapt the requirements discussed in the final MLE report accordingly. MLE participants strongly advocated the further development of EU strategies and policies fostering systemic change in the scientific reward system, including pilot programmes and new instruments for human resources, skills and training.

Where next? A roadmap for Open Science With diverse positions and national initiatives for Open Science at play, the MLE clearly reflected the importance of modular approaches based on monitoring and regular stakeholder exchange. A model roadmap and recommendations for implementing Open Science is described in detail in the MLE report.

However, in order to trigger systemic change in research and research policy, and to make countries fit for the next EU framework funding programme Horizon Europe, several considerations apply:

- The implementation of Open Science needs to be part of the bigger picture, with discussion on the roles and functions of science in society right now, and an agenda and mission for science and innovation based on openness.
- National strategies for the implementation of Open Science are essential to better understand and align the links between Open Science policies and general STI policies. ERA should be the central platform for the development of national OS strategies.
- Champions and role models are needed to foster the uptake of Open Science practices and create a sustainable transition towards more openness.
- Open Science is enhancing knowledge markets and improving innovation. The synergies of scholarly commons (open-access digital repositories) and the commercial exploitation of research outputs require a systematic review and substantial evidence.

Follow-up activities include many presentations of the MLE – nationally and internationally – broad online and offline discussions of the outcomes, and several dedicated events (e.g. presentations in OS-related committees and meetings), as well as a broader dissemination event in Brussels in November 2018. Experts and country delegates alike will ensure the wide dissemination and discussion of the MLE outcomes and thus contribute to European leadership in Open Science in all that it represents.

Thirteen countries participated in the MLE: Armenia, Austria, Belgium, Bulgaria, Croatia, France, Latvia, Lithuania, Moldova, Portugal, Slovenia, Sweden and Switzerland.

For further information:
Reports, minutes and presentations:
https://rio.jrc.ec.europa.eu/policy-support-facility/mle-open-science-altmetrics-and-rewards

Printed in the United States
by Baker & Taylor Publisher Services